T0135624

Gaseous Platinum Clusters –

Versatile Models for Heterogeneous Catalysts

von Diplom-Chemiker Konrad Koszinowski
aus der Freien und Hansestadt Hamburg

von der Fakultät II
– Mathematik und Naturwissenschaften –
der Technischen Universität Berlin
zur Erlangung des akademischen Grades

Doktor der Naturwissenschaften
– Dr. rer. nat. –

genehmigte Dissertation

Promotionsausschuss:

Vorsitzender: Prof. Dr. rer. nat. Siegfried Blechert
Berichter: Prof. Dr. rer. nat. Dr. h.c. Dr. h.c. Helmut Schwarz
Prof. Dr. rer. nat. Jörn Müller

Tag der mündlichen Prüfung: 16. Oktober 2003

Berlin 2003
D 83

Bibliografische Information Der Deutschen Bibliothek

Die Deutsche Bibliothek verzeichnet diese Publikation in der Deutschen
Nationalbibliografie; detaillierte bibliografische Daten sind im Internet über
http://dnb.ddb.de abrufbar.

ISBN 3-8325-0413-3

Logos Verlag Berlin
Comeniushof, Gubener Str. 47,
10243 Berlin
Tel.: +49 030 42 85 10 90
Fax: +49 030 42 85 10 92
INTERNET: http://www.logos-verlag.de

Zusammenfassung

Konrad Koszinowski,

Gaseous Platinum Clusters – Versatile Models for Heterogeneous Catalysts

Kationische Platincluster werden als Modellsysteme für heterogene Katalysatoren vorgeschlagen. Die Reaktivität der Clusterionen lässt sich in der Gasphase mittels Fourier-Transform-Ionencyclotronresonanz-Massenspektrometrie eingehend untersuchen. Die Reaktionen von Pt_m^+-Clustern, $m \leq 5$, mit verschiedenen einfach strukturierten Substraten zeigen bemerkenswerte Ähnlichkeit zu den analogen heterogen katalysierten Prozessen. Dadurch in der Validität des gewählten Modellansatzes bestärkt, wird die Aktivierung von Methan durch Pt_m^+-Cluster im Detail studiert. Die mit den Dehydrierungsreaktionen verbundenen kinetischen Isotopeneffekte, Wasserstoff-Deuterium-Austauschprozesse und energieabhängige Stoßaktivierungsexperimente geben Einblicke in die Potentialhyperflächen der erstaunlich komplexen Systeme Pt_m^+/CH_4. Der wichtigste Unterschied zwischen dem einkernigen Platincarben $PtCH_2^+$ und seinen höheren Homologen $Pt_mCH_2^+$ besteht in der nur für die Cluster möglichen Wechselwirkung von Kohlenstoff mit mehreren Metallzentren. Die dadurch resultierende stärkere Platin-Kohlenstoff-Bindung bewirkt eine drastische Änderung der Reaktivität. Anstelle der für einkerniges $PtCH_2^+$ beobachteten C–N-Bindungsknüpfung führt die Reaktion der $Pt_mCH_2^+$-Cluster mit Ammoniak lediglich zu Carbidkomplexen $Pt_mC(NH_3)^+$. Während der erste Reaktionstyp dem Schlüsselschritt des großtechnischen DEGUSSA-Verfahrens zur platinkatalysierten Blausäure-Darstellung aus CH_4 und NH_3 entspricht, kann das abweichende Verhalten der Cluster als Gasphasen-Analogon zur unerwünschten Verrußung des heterogenen Katalysators angesehen werden. Eine Steuerung der Reaktivität im Gasphasenmodell wird durch den Übergang zu bimetallischen Clustern erreicht. Anders als Pt_2^+ vermitteln $PtCu^+$, $PtAg^+$ und $PtAu^+$ die Kupplung von CH_4 und NH_3 mit hoher Effizienz. Auf der Basis dieser Ergebnisse werden entsprechende bimetallische Katalysatoren zur möglichen Verbesserung des DEGUSSA-Verfahrens vorgeschlagen.

Abstract

Konrad Koszinowski,

Gaseous Platinum Clusters – Versatile Models for Heterogeneous Catalysts

Cationic platinum clusters are proposed as models for heterogeneous catalysts. In the gas-phase, the reactivity of the cluster ions can be thoroughly investigated by means of Fourier-transform ion-cyclotron-resonance mass spectrometry.

The reactions of Pt_m^+ clusters, $m \leq 5$, with several simple substrates exhibit distinct similarities to the analogous processes of heterogeneous catalysts. Encouraged by the thus demonstrated validity of the model approach chosen, methane activation by Pt_m^+ clusters is studied in detail. Kinetic isotope effects associated with dehydrogenation, hydrogen-deuterium exchange processes, and energy-dependent collision-induced dissociation experiments provide insight into the potential-energy surfaces of the surprisingly complex Pt_m^+/CH_4 systems. The main difference between mononuclear platinum carbene $PtCH_2^+$ and the larger homologues $Pt_mCH_2^+$ concerns the interaction of carbon with more than a single metal center in the case of the clusters. The resulting stronger platinum-carbon binding leads to a drastic change in reactivity. Instead of C–N coupling as observed for mononuclear $PtCH_2^+$, the clusters $Pt_mCH_2^+$ yield carbide complexes $Pt_mC(NH_3)^+$ upon reaction with ammonia. Whereas the first type of reactivity corresponds to the key step of the DEGUSSA process for the large-scale synthesis of hydrogen cyanide from CH_4 and NH_3, the deviating behavior of the clusters can be considered as gas-phase analogue of undesired soot formation on the heterogeneous catalyst. In the gas-phase model, reactivity can be controlled if bimetallic clusters are included. Unlike Pt_2^+, the heteronuclear cluster ions $PtCu^+$, $PtAg^+$, and $PtAu^+$ efficiently mediate coupling of CH_4 and NH_3. On the basis of these results, the corresponding bimetallic catalysts are suggested for potential improvement of the DEGUSSA process.

Acknowledgments

First and foremost, I would like to thank my supervisor, Prof. Dr. Dr. h.c. Dr. h.c. Helmut Schwarz, for his trust in my work and for his continuous support. I am deeply indebted to him for most stimulating discussions and the excellent instrumental equipment. Moreover, I wish to thank Prof. Dr. Peter B. Armentrout at the University of Utah, Salt Lake City, for the opportunity to temporarily join his group and to use his guided-ion beam apparatus. Although not directly related to the present thesis, these studies aroused my first interest in transition-metal clusters. I would also like to thank Dr. Martin Bewersdorf, DEGUSSA AG, for enlightening discussions. Furthermore, I am grateful to Prof. Dr. Jörn Müller for willingly being the second examiner.

I would like to express my gratitude to all members of the Schwarz group for their friendly and professional cooperation. Particularly, I wish to thank Dipl.-Chem. Marianne Engeser for the pleasant collaborative use and operation of the FT-ICR mass spectrometer, Dr. Martin Diefenbach for assistance with computational questions, Dr. Ulf Mazurek for providing his ICR Kinetics fitting program, and Dr. Thomas Weiske for maintenance of the instruments and computers. Moreover, I would like to thank Ms Waltraud Zummack and cand.-chem. Philipp Grüne for their help in the field of organic synthesis that was very welcome in the context of two projects independent of the present thesis.

Thanks are also due to the Stiftung Stipendienfonds des Verbandes der Chemischen Industrie for a Kekulé-Stipendium. The Deutsche Forschungsgemeinschaft is acknowledged for further financial support.

My particular thanks go to Dr. Detlef Schröder for his countless suggestions and continuous assistance. I am most grateful for the collaboration with him which I experienced as both highly rewarding and enjoyable.

Table of Contents

Parts of the results of the present thesis have been published in:

1.)   *Probing Cooperative Effects in Bimetallic Clusters: Indications of C–N Coupling of CH4 and NH3 Mediated by the Cluster Ion PtAu+ in the Gas Phase*
       K. Koszinowski, D. Schröder, H. Schwarz, *J. Am. Chem. Soc.* **2003**, *125*, 3676-3677.

2.)   *Reactivity of Small Cationic Platinum Clusters*
       K. Koszinowski, D. Schröder, H. Schwarz, *J. Phys. Chem. A* **2003**, *107*, 4999-5006.

3.)   *Reactions of Platinum-Carbene Clusters PtnCH2+ (n = 1–5) with O2, CH4, NH3, and H2O: Coupling Processes versus Carbide Formation*
       K. Koszinowski, D. Schröder, H. Schwarz, *Organometallics* **2003**, *22*, 3809-3819.

1   Introduction

The twenty-first century poses challenging demands to industrial chemistry. Satisfaction of the ever-increasing need for basic materials has to be brought in line with ecological imperatives requiring reduction of both energy consumption and waste production.[1] Moreover, the limitation of petrochemical feedstocks will necessitate utilization of other resources as substrates for future chemical industry and, thus, demands the development of novel technologies today. Clearly, the search for higher efficiencies, higher selectivities, and new types of reactivity will continue to center upon transition-metal catalysis as an essential component of modern chemistry.[1-5]

The unique potential transition metals offer to synthetic applications mainly results from their unsaturated electronic valence shells. In general, the availability of empty orbitals allows transition-metal centers to easily switch between different oxidation states and coordination numbers such that they are ideally suited for participation in catalytic cycles. With the *d* block elements of the periodic table, nature provides a large pool of possible candidates for the realization of a specific catalytic conversion. However, experience shows that not all transition-metal elements have equal catalytic activities. Instead, some transition metals catalyze a manifold of different reactions, whereas the catalytic activities of others are limited to just a few reaction types or even almost completely absent.

One of the most versatile metals in catalysis certainly is platinum. First applied in heterogeneous catalysis two centuries ago,[6,7] its utilization has experienced a tremendous rise to date. *Inter alia*, heterogeneous platinum catalysts are crucially involved in hydrocracking and platforming in the petrol industry,[8,9] in dehydrogenation and hydrogenation reactions of organic substrates,[10-15] in the Ostwald process synthesizing nitric oxide,[16] in the DEGUSSA and Andrussov processes for the large-scale synthesis of hydrogen cyanide,[17-22] and last but not least in the decomposition of carbon monoxide and nitrous oxide in automotive catalytic converters.[23-25] Despite their enormous economic relevance, the current understanding of these reactions is relatively poor because of the difficulties inherent in mechanistic investigations of heterogeneous systems. Although there has been recent progress in the *in-situ* characterization of bulk catalysts, the insights gained from these studies in most cases do not suffice for rational catalyst improvement.[26,27]

Gas-Phase Models of Heterogeneous Catalysts. An alternative and complementary strategy does not approach the complex heterogeneous catalysts themselves, but considers model systems that are simple enough for thorough mechanistic elucidation while still featuring the essential characteristics of the real catalytic process. The most drastic simplification conceivable focuses only on the intrinsic reactivity of the catalytically active transition metal itself and excludes all other constituents of the heterogeneous system like neighbor atoms, ligands, ad-atoms, or counterions. Such a model is realized by gaseous transition-metal ions whose chemical behavior can be investigated in-depth by advanced mass spectrometric methods.[28] Extensive studies of mono-cationic transition-metal ions in general and platinum ions Pt^+ in particular have demonstrated remarkable similarities between the reactivities of these model systems and bulk catalysts. For instance, gaseous platinum cations Pt^+ readily induce dehydrogenation of hydrocarbons and, thus, mirror the reactivity of platinum surfaces used in petroleum processing.[29-31] Notably, Pt^+ even succeeds in the activation of methane as the most inert hydrocarbon.[31-37] This reaction is especially intriguing, because its counterpart in heterogeneous catalysis would open a highly desired avenue to the utilization of natural gas as raw material for chemical industry.[3,38] A well-established example of such a methane functionalization is given by the DEGUSSA process with the Pt-catalyzed coupling of methane and ammonia in the key step. For this reaction, an elaborate gas-phase model, again based on Pt^+, could be developed.[39,40] Similarly, oxidation processes like those occurring in automotive catalytic converters were mimicked by the reactions of the gaseous ions PtO^+ and PtO_2^+.[41]

Considering these parallels in reactivity, the approach starting from Pt^+ as model for heterogeneous platinum catalysts indeed appears reasonable. For achieving still closer analogy, however, refinement of the model system obviously is necessary. A straightforward modification replaces atomic Pt^+ by small cluster ions Pt_m^+. Distribution of the unit Coulomb charge over the m platinum atoms significantly reduces the charge density of the individual atoms and thereby almost approaches electroneutrality of solid-state catalysts. With increasing cluster size m, the degree of aggregation takes an intermediate position between the isolated atom and the bulk limit as well. This transition does not only affect the system's geometric features but also dramatically alters its electronic configuration.[42] As an associative phenomenon, emergence of the metallic state requires a minimum size which is supposed to fall within the range of medium-sized clusters. For smaller metal clusters, the electronic properties often display pronounced non-monotonic variations that are reflected in size-specific reactivities.[43] Such effects are of prime importance in the context of heterogeneous catalysis where the active sites are commonly believed to correspond to point defects or edge dislocations, *i.e.*, structures resembling small isolated clusters.[44] A prominent example of size-specific reactivity is given by gold as platinum's right-handed neighbor in the periodic table. Whereas small gold clusters and nanofilms exhibit a distinct catalytic activity with respect to oxidation reactions, similar processes occur neither for single gold atoms nor for the metal in the solid state.[45-51]

Besides homonuclear clusters Pt_m^+, heteronuclear clusters may represent interesting models for heterogeneous catalysts as well. Quite often, for instance, the catalytically active transition metal is not used in bulk quantities but loaded onto a support such as alumina or silica.[52-54] The influence of these supports on the catalytic activity currently is not well-understood. Provided gas-phase models like $PtAl_2O_3^+$ or $PtSiO_2^+$ could be realized, a comparison between the reactivities of such species and bare Pt^+ might be helpful in this regard. Heteronuclear clusters can also be used to mimic bimetallic catalysts. Addition of a second transition metal to the heterogeneous catalyst is an important means for fine-tuning and optimization of catalyst performance.[55] However, this methodology has essentially remained empirical to date (for an exception, see Chapter 7) such that the development of more efficient bimetallic systems is rather laborious. In view of the particular complexity of bimetallic heterogeneous catalysts, simple models appear extremely valuable for mechanistic elucidation. Again, gas-phase studies promise to reveal the intrinsic reactivities of the bimetallic systems as a starting point for possible further investigations. In addition, gas-phase techniques might prove useful for screening the reactivities of different combinations of transition metals prior to the more demanding exploration under heterogeneous conditions.

From the potential applications of platinum clusters as gas-phase model systems outlined above, only those relying on homonuclear clusters have been considered so far. Research on cationic platinum clusters Pt_m^+ has concentrated on probing their reactivities toward hydrogen and hydrocarbons. Pt_m^+ clusters were found to adsorb hydrogen[56] and readily induce dehydrogenation of several alkanes,[29,30] ethylene,[31] acetylene,[31] and arenes.[30] Like Pt^+, the homologous clusters also react with methane yielding $Pt_mCH_2^+$; the efficiencies of these processes display an interesting dependence on the cluster size m.[43,57] Analogous reactions occur for neutral platinum clusters,[43,58] which also dehydrogenate hexane and benzene,[59] and for the corresponding anionic clusters.[57] Thus, the clusters' distinct capability to activate the smallest hydrocarbon is by and large independent of the charge state, thereby further encouraging their usage as model systems for heterogeneous catalysis. The few remaining reactions studied so far involve small anionic platinum clusters Pt_m^-. Besides formation of association products and cluster degradation upon reaction with nitrogen, oxygen, carbon monoxide, and carbon dioxide, oxygen transfer occurs in the reaction with nitrous oxide.[60,61] The resulting products Pt_mO^- oxidize carbon monoxide under regeneration of Pt_m^- such that complete catalytic cycles in the gas phase emerge.[62,63] The analogy to processes in automotive catalytic converters once more emphasizes the considerable potential platinum clusters offer as models for heterogeneous catalysts. However, with regard to platinum's eminent role in catalysis, the chemistry especially of the cationic platinum clusters seems to be much less explored than desirable.

Scope of the Present Thesis. The present thesis aims to bridge this gap and seeks to systematically investigate the reactivities of cationic platinum clusters in the gas phase. One of its major directions concerns the reactions of Pt_m^+ clusters, $m \leq 5$, with small inorganic substrates (Chapter 3). Knowledge of these processes is indispensable for a chemical characterization of the clusters and, in addition, is needed to address a wider scope of reactions relevant to heterogeneous catalysis. Given the particular importance of methane activation, this process deserves special attention. Besides a detailed exploration of the dehydrogenation reactions (Chapter 4), the reactivities of the thus formed products $Pt_mCH_2^+$ and Pt_mC^+, $m \leq 5$, are in the center of the focus. In analogy to the previous work on mononuclear platinum, their reactions with oxygen and the prototypical nucleophilic substrates water and ammonia are probed (Chapter 5). Among these, the reactions with ammonia and the question of C–N coupling appear particularly attractive as they might expand the previous gas-phase model of the DEGUSSA process for the synthesis of hydrogen cyanide. Moreover, some heteronuclear platinum clusters are included in the present work. First, the possibility of developing model systems for supported platinum is investigated (Chapter 6). Further, the reactivities of the bimetallic cluster ions $Pt_mAu_n^+$, $m + n \leq 4$, as well as $PtCu^+$ and $PtAg^+$ are studied (Chapters 7 - 9). Again, the main focus lies on their chemical behavior with regard to the DEGUSSA process.

Throughout the thesis, the reactivities of the gaseous clusters are compared with those of heterogeneous catalysts. The degree of agreement achieved can serve as a measure of the adequacy of small cationic platinum clusters as model systems for the corresponding solid-state catalysts. Provided good accordance does exist, the mechanistic implications of the gas-phase model may be used for a possibly new interpretation of reactions occurring under heterogeneous conditions. A progress in understanding could ultimately lead to rational catalyst improvement. The last issue of this agenda, however, is far beyond the scope of the present work.

2   Experimental Methods

The experiments are performed by means of Fourier-Transform Ion-Cyclotron Resonance mass spectrometry (FT-ICR-MS).[64-67] Thanks to the ultra-high mass-resolving power of this technique and its particular suitability for MS/MS experiments, FT-ICR mass spectrometry proves to be the method of choice for reactivity studies of mass-selected platinum clusters Pt_m^+ in the gas phase. A typical experiment starts with laser vaporization/ionization of the metal in the external ion source (Figure 2.1). Co-expansion with helium into a vacuum produces cold metal-cluster ions that are transferred into the analyzer cell of a Bruker Spectrospin CMS 47X FT-ICR mass spectrometer.[68,69] There, the cluster ions are exposed to the neutral reactants and subjected to series of isolation, acceleration, and detection events. In the following, the single steps of the experiment are described in detail with special consideration of the problems inherent in the treatment of cluster ions.

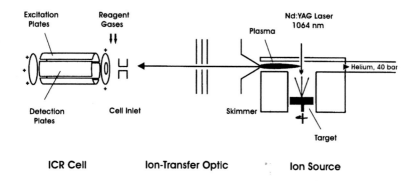

Figure 2.1. Schematic view of the FT-ICR mass spectrometer with the cluster-ion source.

2.1   Cluster-Ion Generation

Cluster-ion generation takes place in a modified[70] Smalley-type[71] cluster-ion source developed by Niedner-Schatteburg, Bondybey, and co-workers.[72] The fundamental of a pulsed Nd:YAG laser (Spectron systems, 0.170 J maximum output energy per pulse, 5 Hz repetition rate) is focused onto a rotating metal target. The thus generated hot metal plasma is supposed to consist of neutral and charged metal atoms. Formation of neutral and ionic clusters is accomplished by a synchronized helium gas-pulse (static pressure varied up to 40 bar) that consumes the excess energy released upon particle condensation. Further cooling is achieved by supersonic expansion which efficiently converts a large fraction of the clusters' internal energy into the directed translational motion by collisional coupling. The remaining small portions of internal energy correspond to low effective rotational and vibrational temperatures and thus permit the preparation of weakly-bound systems such as van-der-Waals complexes. To probe the cooling efficiency of the present experimental set-up, an iron plasma is expanded in a helium jet seeded with nitrogen. Thereby, formation of $Fe(N_2)_x^+$ ions up to $x = 4$ is observed. Given the rather low bond-dissociation energies of these complexes $(D_0((N_2)_3Fe^+–N_2) = 0.56 \pm 0.04$ eV),[73] the cluster ions generated cannot have substantially elevated internal energies because dissociation would have occurred otherwise. Whereas the higher stability of the Pt_m^+ clusters produced prevents their dissociation in any case, the absence of excess internal energy may be assumed in analogy to the $Fe(N_2)_x^+$ system.

After supersonic expansion, the molecular beam passes a skimmer and a system of electrostatic potentials and lenses which guide its cationic components into the analyzer cell while the neutral particles are removed by differential pumping. The final abundance of a given cluster size crucially depends on a careful adjustment of the different electro-optical devices. Here, the electric potential between the cluster source and the inlet into the analyzer cell[74] proves to be especially important, as it controls the energy of the clusters entering the ICR cell. With an insufficient translational energy, the cluster ions cannot enter the cell fast enough.[75] In contrast, too energetic ions are no longer trapped within the analyzer cell (see below). Obviously, the potential applied between cluster source and inlet increases or decreases the translational energies E_{trans} of all mono-cationic cluster ions by the same amount and, thus, does not inflict any mass discrimination by itself. However, a mass discrimination does arise from the constant velocity v of all cluster ions imposed by the entraining jet stream at the outlet of the cluster source. According to the elementary relation $E_{trans} = \frac{1}{2} m v^2$, ions with larger masses m have higher translational energies than those with smaller masses. Therefore, optimization of E_{trans} should require increasingly repulsive potentials as a function of cluster size. Indeed, this behavior is observed *in praxi*. The remaining more than 30 variables are closely interrelated, such that the arising complexity of the system only permits an essentially empirical approach.

Besides the adjustment of the ion-transfer optics, the conditions in the ion source are of equal importance for obtaining reasonably high cluster abundances. Here, the static pressure of the helium, the delay between laser shot and helium pulse, and the duration of the latter constitute the most sensitive parameters. Again, the variables are closely interdependent such that changes in one parameter can be partially compensated by altering the others. For different cluster sizes, the conditions have to be changed more or less significantly without any clear rules being evident. Usually, a satisfactory setting found keeps working for several days before major adjustment becomes necessary. With respect to the finding of a new setting, starting from mononuclear Pt^+ has proven to be most favorable. In contrast to the cluster ions, Pt^+ is generated in high abundances over a wide range of parameter settings and can thus be found rather easily. After optimization, the relative potential of the cluster source and the potential of the flight tube[76] as the two most sensitive parameters of the transfer optics are subsequently scanned while observing the m/z range expected for Pt_2^+ in the high resolution mode (see Section 2.3). In most cases, this procedure finds conditions that produce Pt_2^+ in detectable amounts. Once platinum-dimer ions are formed, optimization of their abundance usually is achieved by repeated scanning of all parameters of the ion source and the transfer-optics in a relatively straightforward manner. The instrumental settings adjusted to the generation of Pt_2^+ typically also yield larger clusters Pt_m^+ in smaller amounts that can serve as initial points for an analogous optimization. For all of the clusters, however, the best abundances attained still are at least one order of magnitude smaller than that of atomic Pt^+; similar situations are also found for other transition metals investigated. Consequently, extensive data accumulation (up to 2000 scans) is necessary for achieving reasonable signal-to-noise ratios.

2.2   Ion Motion in the ICR Cell

After entering the ICR cell, the ions are trapped therein and stored at ultra-high vacuum conditions for theoretically any time. Usually, the ions generated in a sequence of six laser shots, each followed by a helium pulse, are accumulated prior to further manipulation. Confinement of the ions in the *xy* plane is achieved by location of the cell in the spatially uniform field within a super-conducting magnet. Ions of charge q and mass m moving with velocity v_{xy} in the *xy* plane perpendicular to the magnetic flux density B_z are bent to a circular path by operation of the Lorentz force F_L. For a given flux density B_z, the resulting (angular) cyclotron frequency ω_c only depends on the ions' charge to mass ratio as easily seen by identification of F_L with the centripetal force F_c (eq. 2.1a, where r is the radius of the ions' circular motion). In the present case, $B_z = 7.05$ T corresponds to cyclotron frequencies about 1 MHz for small mono-cationic platinum clusters.[77]

$$F_L = F_c \iff q \, v_{xy} B = m \, v_{xy}^2 \, r^{-1} \implies \omega_c = v_{xy} \, r^{-1} = q \, B \, m^{-1} \qquad (2.1a)$$

Trapping of the ions along the z axis is ensured by application of repulsive electrostatic potentials $V_{trap} \approx 1.5$ V at both end caps of the ICR cell.[78] For the given cylindrical geometry, the following dependence results for the trapping potential Φ (eq. 2.2),

$$\Phi = V_{trap} \, [\gamma + \tfrac{1}{2} \alpha a^{-2} (2 z^2 - r^2)] \qquad (2.2)$$

where γ, α, and a are parameters describing the geometry of the trap.[67] Obviously, the axial confinement introduces a radial force which opposes the Lorentz force, thus modifying eq. 2.1a and yielding eq. 2.1b.

$$m \, \omega^2 r = q \, B \, \omega \, r - q \, V_{trap} \, \alpha a^{-2} r \qquad (2.1b)$$

The solutions of this quadratic equation in ω give the perturbed cyclotron frequency ω_+ (eq. 2.3a) and the so-called magnetron frequency ω_- (eq. 2.3b) which belongs to a circular motion superimposed to the cyclotron motion.

$$\omega_+ = \tfrac{1}{2} \omega_c + [(\tfrac{1}{2} \omega_c)^2 - \tfrac{1}{2} \omega_z^2]^{\frac{1}{2}} \qquad (2.3a)$$

$$\omega_- = \tfrac{1}{2} \omega_c - [(\tfrac{1}{2} \omega_c)^2 - \tfrac{1}{2} \omega_z^2]^{\frac{1}{2}} \qquad (2.3b)$$

Both motions are superimposed by an oscillation in the z axis with a frequency ω_z (eq. 2.3c).

$$\omega_z = [2 q\, m^{-1}\, V_{trap}\, \alpha\, a^{-2}]^{\frac{1}{2}} \tag{2.3c}$$

Usually, the relation $\omega_+ \gg \omega_z \gg \omega_-$ holds true such that only the perturbed cyclotron frequency ω_+ is recorded.

2.3   Ion Detection

Detection of the ions relies on the measurement of the image current induced by the circulating ions in a pair of detector plates. For obtaining a detectable signal, the originally statistical distribution of the ions' phase angles must be transformed into a coherent one. This transformation can be accomplished by resonant excitation of the ions. If the frequency of an external, spatially uniform electric field matches ω_+, the ions are coherently accelerated to a higher orbit. Besides establishing coherency, this increase of the ions' orbit also enhances the image current and, thereby, the signal intensity. Whereas simple ICR mass spectrometry sequentially scans the m/z range of interest, the FT-ICR technique excites all ions with different m/z ratio simultaneously and records the evolution of the resulting signal in time. A Fourier transformation from the time to the frequency domain provides a conventional mass spectrum.[79]

Two different acquisition modes can be distinguished. In the case of broadband excitation and detection, a wide m/z range is covered in order to ideally record all ions present in the cell. Whereas there is no theoretical upper limit for the detectable m/z range, a lower limit of $m/z > 17$ is imposed by the response time of the instrument's electronics. Usually, however, not the whole accessible m/z range is covered but only that part where ions of interest are expected. For instance, it was carefully checked that the reactions of Pt_m^+ clusters with the substrates H_2, O_2, H_2O, N_2O, and CH_4 at thermal energies do not yield any products with masses smaller than that of Pt^+; the kinetic investigations therefore can be restricted to a m/z range from 170 to 3000 for maximizing the spectral resolution and minimizing the acquisition of bare noise. However, the recorded m/z window should not be too narrow either, because otherwise ion excitation (and detection) may suffer from mass discrimination. Because only the center of the excitation window exhibits a constant amplitude, ions lying in the edge regions experience differing electric fields such that their signal intensities can no longer be correlated with their real abundances. Ion acquisition in these edge regions of the excitation range cannot be avoided if ions with very small and large m/z values are present simultaneously. This unfavorable situation may arise in the case of charge transfer from platinum-cluster ions to small neutral substrates (see Section 3.2 for an example) and can impair the quantitative analysis of these reactions in kinetic studies.

The second acquisition mode only scans a very narrow m/z window. The much longer detection time thus available for a small m/z range enhances the mass resolution achievable up to $m/\Delta m > 10^6$ for sufficiently low pressures. Application of somewhat higher pressures as required for reactivity studies increases the collisional dampening of the signal in the time domain and thereby leads to slightly lower mass resolutions. Nonetheless, these are in general still sufficient to unambiguously establish the ions' elemental compositions by determining their exact masses (see Figure 9.1 for an example).

2.4   Ion Selection

Ion excitation is not only indispensable for ion detection but also highly useful in ion selection. If an unwanted ion is accelerated to an orbit larger than the cell geometry, it hits the wall and is neutralized. Such a means for removal of unwanted ions and thus ion selection is particularly warranted for cluster studies, where initially ions of several cluster sizes and their products are present simultaneously. In the case of platinum, a further difficulty arises from the element's isotope pattern.[80] In reactivity studies, the broad isotopic distributions arising for Pt_m^+ clusters (Figure 2.2a) can severely complicate the analysis of the products, especially if isotopically labeled compounds are used. Whereas Marshall and co-workers handled this problem by the application of polynomial methods,[30] a different approach is chosen in the present work. Here, mass selection is not restricted to a specific cluster size m, but extended to the isolation of a single isotopomer, namely the formal $^{195}Pt_m^+$ isotopomer (Figure 2.2b).[81]

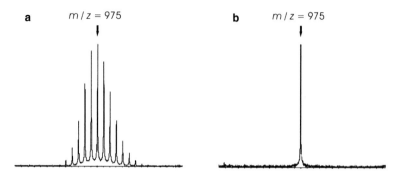

Figure 2.2. Mass range m/z = 962 - 988 (a) prior to and (b) after selection of the Pt_5^+ ion with m/z = 975 and its thermalization.

Broadband and Single-Ion Ejection Modes. Similarly to ion detection, both broadband and single-ion acceleration can be used for ion ejection.[82] Again, it has to be taken into account that the excitation pulses are not strictly rectangular but have more or less broad flanks. If these flanks reach to the ion to be selected, it is subject to unintended excitation, in the following referred to as off-resonant excitation. In such a case, the chemical processes observed no longer correspond to thermal reactivity because the ions have elevated kinetic energies. Clearly, occurrence of off-resonant excitation is particularly likely for the isolation of a single Pt_m^+ isotopomer. According to eq. 2.1a, a mass difference $\Delta m/z = 1$ for two ions of $m/z \approx 1000$, like Pt_5^+, leads to a frequency difference $\Delta\omega_c$ that is smaller by a factor of 10 than that resulting for two neighboring ions of $m/z \approx 100$. A possible means for the exclusion or reduction of off-resonant excitation is the application of sufficiently sharp and well-defined ejection pulses. However, strict localization in the frequency domain necessarily corresponds to a large temporal extension of the electric field $\mathscr{E}$, eq. 2.4 (valid for single-ion excitation), where t_{exc} is the excitation time.[67]

$$\mathscr{E}(\omega_c/2\pi) = \mathscr{E}_0 \frac{\sin(\omega_c \, t_{exc})}{\omega_c} \tag{2.4}$$

If, like in the case of $^{195}Pt_m^+$ isolation, numerous ions are to be ejected, application of extremely sharp pulses takes considerable time. Yet, exceedingly long isolation procedures cannot be tolerated in the course of reactivity studies because otherwise the abundance of the reactant ion is severely diminished because of product formation. Ideally, the excitation time t_{exc} could be varied as a function of m/z such that very accurate ejection pulses could be used to eject the direct neighbors of the $^{195}Pt_m^+$ ion, whereas broader pulses would suffice for the removal of more remote ions. However, the FERETS (Front End Resolution Enhancement by using Tailored Frequency Sweeps)[83] technique utilized in the present experiments does not include this option but only has few global parameters for controlling the instrument's operation in ion ejection. Therefore, a trade-off between accuracy and time consumption of the ejection procedure has to be accepted.[84]

Experimental Probes for Off-Resonant Excitation. For the ion-ejection settings used throughout the present experiments, the problem of off-resonant excitation is addressed systematically in the case of trinuclear platinum clusters. As a probe, the reaction of the mass selected $Pt_3CH_2^+$ ion with $m/z = 599$ with argon is monitored.[85] Whereas no reaction occurs at thermal energies, dehydrogenation according to reaction 2.5 takes place at elevated energies and, thus, indicates an enhanced kinetic energy of the reactant ion (see Section 4.4).

$$Pt_3CH_2^+ + Ar + \Delta E \quad \rightarrow \quad Pt_3C^+ + H_2 + Ar \qquad (2.5)$$

The effect of broadband ejection is investigated by application of an ion-ejection window above $m/z = 599$ and variation of its lower boundary m_{lim}/z (Figure 2.3, inset). No reaction takes place for $m_{lim}/z \geq 750$. Further extension of the ejection window towards $Pt_3CH_2^+$ induces rapidly increasing dehydrogenation. Obviously, the flank of the ejection pulse spans a range $m/z \approx 100$ wherein off-resonant excitation occurs. Despite tuning the instrument to the generation of rather well-defined ejection pulses, the shape of the broadband windows apparently still is rather poor.

A similar method is used for probing off-resonant excitation resulting from single-ion ejection. Whereas no effect is observed for ejection pulses ≥ 3 mass units away from $Pt_3CH_2^+$ with $m/z = 599$, dehydrogenation begins to occur upon ejection of $m/z = 601$ and predominates for a pulse applied on $m/z = 600$ (Figure 2.4). Clearly, the single-ion ejection pulses are not narrow enough to avoid off-resonant acceleration either. As a further improvement of the ejection accuracy is prohibited by temporal requirements (see above), alternative means are needed for the suppression of unwanted ion excitation.

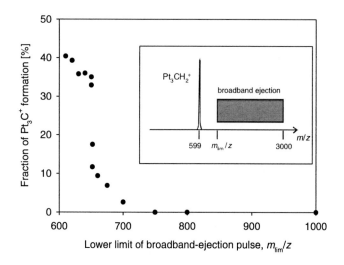

Figure 2.3. Off-resonant excitation of $Pt_3CH_2^+$ (m/z = 599) as a function of the broadband ejection window. Whereas the upper limit of the ejection window is held constant at m/z = 3000, the lower limit m_{lim}/z is varied between 610 and 1000 as shown in the inset. Off-resonant excitation is probed by the formation of Pt_3C^+ upon collision with argon gas.

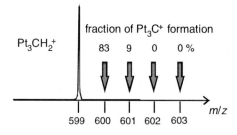

Figure 2.4. Off-resonant excitation of $Pt_3CH_2^+$ (m/z = 599) as a function of single-ion ejection pulses applied at m/z = 600 - 603. Off-resonant excitation is probed by the formation of Pt_3C^+ upon collision with argon gas.

Ion Thermalization. A simple and efficient way for the deactivation of kinetically excited ions is their collision with a buffer gas such as argon. With every single collision decreasing the ion's kinetic energy, complete thermalization is achieved provided sufficient collisions. Because high pressures also enhance the collisional dampening of the ions' signals during their detection, however, a permanent application of the buffer gas usually is less favorable. Thus, in the present experiments, argon is pulsed in after mass selection of Pt_m^+ and pumped off prior to ion detection. Typically, the pressure reaches a maximum of 10^{-5} mbar and decreases to its initial value within 1 - 2 s. The success of this thermalization is probed by means of the reaction of the isolated Pt_m^+ clusters with CH_4, m = 2 - 5. Whereas mainly double dehydrogenation affording Pt_mC^+ is observed without thermalization (reaction 2.6a), solely formation of $Pt_mCH_2^+$ takes place if argon is pulsed in after the isolation procedure (reaction 2.6b). The latter reaction corresponds to the process known for thermal Pt_m^+ clusters and thereby demonstrates the effectiveness of the thermalization applied.[57]

$$Pt_m^+ + CH_4 + \Delta E \quad \rightarrow \quad Pt_mC^+ + 2\,H_2 \qquad\qquad (2.6a)$$

$$Pt_m^+ + CH_4 \quad \rightarrow \quad Pt_mCH_2^+ + H_2 \qquad\qquad (2.6b)$$

Another probe for thermal *versus* non-thermal behavior is given by the rate of a specific reaction. For kinetically excited ions, the reaction rate should change with time because of gradual radiative or collisional relaxation. Similarly, for an insufficient thermalization, differences in the observed rate constants would be expected after application of further argon pulses. The absence of such phenomena in the present case once more proves complete thermalization of the platinum-cluster ions investigated. Moreover, the observation of constant reaction rates until high degrees of conversion excludes significant electronic excitation of the reactant cluster-ions as well.

2.5   Collision-Induced Dissociation

Ion excitation is also needed for collision-induced dissociation (CID) experiments. These form a versatile means to probe an ion's structure by analysis of its fragmentation products. The usual experimental sequence starts by preparation and mass selection of the ion of interest. After its acceleration by single-ion excitation, the ion is collided with a buffer gas such that fragmentation occurs like in the case of the $Pt_3CH_2^+$ clusters upon off-resonant excitation. Finally, the ionic fragmentation products are detected in the broad-band mode. Variation of the collision energy can be achieved by changing the excitation energy. High excitation and, thus, collision energies give rise to extensive fragmentation of the parent ion because the energy is sufficient for breaking several bonds. In contrast, if the collision energy is low enough, only the energetically least demanding process occurs, *i.e.*, fission of the weakest bond.[86] This relation quite often allows structural assignments although the possibility of complicating rearrangement reactions has to be considered.

In the present work, argon is used as collision gas.[87] The collision energy usually is varied from the onset of dissociation up to > 50 % fragmentation. Note that also for these experiments, isolation of a single isotopomer is essential because otherwise ill-defined fragmentations of the neighboring isotopomers would result from off-resonant excitation.

Besides the qualitative analysis of CID experiments, also their quantitative evaluation is possible. Such an approach promises to determine the energetic threshold of a dissociation reaction, and, thus, the corresponding binding energy.[86] However, the reliability of energy-resolved FT-ICR experiments has been questioned as at least in some early studies, numerically correct results were derived on the basis of wrong assumptions. Although there has been progress in the physical description of the ion-excitation process meanwhile, some inconsistencies in the literature remain to date.[88,89] The present experiments rely on the work of Sievers *et al.* who not only performed a thorough theoretical analysis but also probed the validity of their approach experimentally.[89,90] According to their findings, the ion-cyclotron energy E_c is given by

$$E_c = \frac{\beta^2 e^2 V_{PP}^2 t_{exc}^2}{128 \, r^2 m} \qquad (2.7)$$

with the geometrical factor of the ICR cell $\beta = 0.83$ in the present case,[91] the elementary charge e (for mono-charged ions), the peak-to-peak voltage of the excitation plates $V_{PP} = 8.4$ V,[92] the radius of the cell $r = 0.03$ m, and the ion's mass m. E_c is varied by changing the excitation time t_{exc}. Conversion of the collision energy from the laboratory to the center-of-mass (CM) frame follows

$$E_{CM} = \frac{m_{Ar}}{m + m_{Ar}} E_{lab} \qquad (2.8)$$

with $E_{lab} = E_c$. Eq. 2.8 holds strictly for the single-collision limit only because otherwise an ill-defined amount of energy is transferred in additional collisions. To minimize multiple collisions, the pressure applied is kept as low as $p(Ar) \approx 1.5 \times 10^{-8}$ mbar and the collision time is limited to 0.5 s in all threshold experiments.[93]

2.6 Ion-Molecule Reactions

After preparation, mass selection, and thermalization, the cationic platinum clusters are exposed to the neutral reactants added to the cell via one of two leak valves.[94] Typically applied pressures between 5×10^{-9} and 10^{-6} mbar are determined by means of a calibrated[95] Bayard-Alpard ion gauge under acknowledgment of the substrates' relative sensitivities.[96,97] Compared to the ionic reactant A^+, the neutral substrate B is present in large excess such that pseudo first-order reactions evolve,

$$\frac{d[A^+]}{dt} = -k\,[A^+][B] \cong -k_{obs}\,[A^+], \text{ with } k_{obs} = k\,[B] \tag{2.9}$$

where k is the true bimolecular and k_{obs} the apparent pseudo-unimolecular rate constant. Recording a temporal profile of the reaction (usually, spectra are acquired for 5 - 10 different reaction times) gives the reactant ion's decline whose negative slope in a semi-logarithmic plot corresponds to k_{obs}. The good quality of the linear regression achieved in most cases reflects a rather high precision achieved for k_{obs}. However, derivation of the relevant bimolecular rate constants k relies on knowledge of the pressure as well. Because of the limited accuracy of ion gauges with respect to the determination of absolute pressures,[96,97] the final uncertainties associated with the absolute rate constants k amount to $\pm$ 30 % as shown by systematic investigations.[98] For the neutral reactants ammonia, methylamine, and water, their unfavorable pumping characteristics and adsorption behavior also render the maintenance of a constant pressure throughout the whole experiment difficult. To include this additional source of uncertainty, the error bars of the corresponding rate constants are raised to $\pm$ 50 %. In the case of water, pressure-dependent studies are performed in order to account for this substrate's inevitable presence in the background of the instrument's high-vacuum system.

Kinetic Analysis of Consecutive Reactions I: Direct Approach. For the quantitative analysis of consecutive reactions, two different approaches are applied. First, the composite reaction scheme can be reduced to a set of primary reactions by subsequent mass-selection of all intermediates. This procedure represents the most straightforward way to tackle the problem and therefore is in principle preferable though also rather time-consuming, particularly for inefficient primary reactions. The time demand can be substantially reduced if the neutral substrates are not only added permanently via leak valves but also temporarily using pulse valves. Pulsing-in the reactants leads to a rapid build-up of relatively high pressures such that the intermediates are readily formed. Because the substrates are pumped off within 1 - 5 s, they do not interfere with the consecutive reactions if an appropriate time delay is applied. Therefore, this option is ideally suited for the combination of different neutral reactants, whereas simultaneous leaking-in of two different substrates does not allow an unambiguous determination of their partial pressures. For instance, pulsing-in methane after mass selection and thermalization of Pt_m^+ efficiently affords the carbene clusters $Pt_mCH_2^+$ whose reactivities toward leaked-in substrates such as ammonia or oxygen can be accurately probed subsequently. If the thermalization is left out, the kinetically excited Pt_m^+ ions mainly yield carbide clusters Pt_mC^+ according to reaction 2.6a.[99,100] After ejection of co-generated $Pt_mCH_2^+$ and thermalization, the reactivities of these species can be characterized as well. If the number of added substrates is further increased, even multi-step syntheses can be realized in the gas phase. An example for such an approach is given in Section 6.2.

Kinetic Analysis of Consecutive Reactions II: Modeling. The second strategy for the investigation of consecutive processes treats all reactions simultaneously and seeks a numerical solution for the overall reaction kinetics. This method starts with the postulation of a chemically plausible kinetic scheme which is fitted to the experimental product evolution. Whereas this approach works well for relatively simple reaction systems,[101] ambiguities can arise for more complex kinetic schemes. These may have so many different adjustable parameters that the product evolution observed might be reproduced within the experimental uncertainties irrespective of the model's physical validity. In this case, the agreement between experimental and predicted product distribution can obviously no longer serve as criterion for the quality of the model. Despite this *caveat*, the usage of the modeling approach may be necessary if not all of the intermediates can be isolated and studied separately. For instance, a combination of slow primary but rapid consecutive reactions may prevent substantial accumulation of the reaction intermediate such that its selection might not be practical. A second obstacle to isolation of an intermediate may arise from the possibility of unwanted off-resonant excitation which is likely to occur if the masses of the different species are very similar (see Section 2.4). Such a case is met for the isotopic exchange reactions between $Pt_mCH_2^+$ clusters and D_2 (see Section 4.3). Mass selection of the intermediate Pt_mCHD^+ ions would require ejection of both the ionic reactants $Pt_mCH_2^+$ and products $Pt_mCD_2^+$ which leads to off-resonant excitation of Pt_mCHD^+ itself. Subsequent thermalization by pulsing-in argon is not feasible either, because the resulting time delay would again allow formation of $Pt_mCD_2^+$. Therefore, the complete reaction cascades starting from $Pt_mCH_2^+$ are recorded and analyzed by means of numerical routines which yield the rate constants of the individual reactions.[102,103] To test the sensitivity of the kinetic models derived in this way, each rate constant k_i is varied stepwise and held constant while all other k values are kept free for adjustment to experiment. This procedure is repeated until the restricted model can no longer reproduce the experimental evolution of the ionic species. The associated acceptable ranges for k_i are taken as a measure for their uncertainties within the postulated kinetic model.

Efficiencies of Ion-Molecule Reactions in the Gas Phase. With respect to the absolute rate constants, clearly a reference value is needed for comparison. Such a value is given by the theoretical prediction of the maximum rate possible for a certain reaction, the so-called collision rate. Its calculation relies on the assumption that both reaction partners have to come into close contact for a reaction to occur. In general, surpassing of the rotational barrier as the highest point of the effective inter-particle potential is considered necessary for a sufficient spatial approach. For the interaction between an ion of charge q and a neutral with a polarizability volume $\tilde{\alpha}$, the collision rate is given by

$$k_{LGS} = \sqrt{\frac{\pi \tilde{\alpha} q^2}{\varepsilon_0 \mu}}, \quad \mu = \frac{m_{ion}\, m_{neutral}}{m_{ion} + m_{neutral}} \qquad (2.10)$$

with the vacuum permittivity ε_0 and the reduced mass μ in SI units. First proposed by Langevin and later on taken up by Gioumousis and Stevenson, eq. 2.10 is usually referred to as Langevin-Gioumousis-Stevenson (LGS) model.[104,105] For neutrals without large permanent dipole moments, its prediction in general agrees with experiment reasonably well whereas considerable discrepancies emerge in the case of polar neutral substrates. To account for these cases as well, several modifications of the LGS model have been proposed of which the capture theory developed by Su is the most wide-spread.[106,107] In the present work, collision frequencies k_{cap} calculated according to capture theory are used for the determination of reaction efficiencies $\varphi \equiv k \, / \, k_{cap}$.

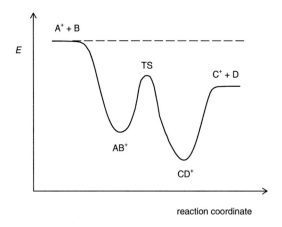

Figure 2.5. Double-minimum potential for a typical ion-molecule reaction.

The efficiency of a given reaction may also be correlated with the system' potential energy surface. Because the interaction between an ion A^+ and a neutral B always is attractive, there are no thermochemical constraints with respect to the formation of an ion-molecule complex AB^+ in the first step. In marked contrast to the condensed phase, the gas phase does not provide a heat bath such that the energy released from complex formation is not dissipated but remains as rotational and vibrational excitation in the encounter complex. For further proceeding of the reaction, the adduct AB^+ has to undergo a rearrangement which is associated with a barrier in the system's potential-energy surface (Figure 2.5). Clearly, surpassing of this barrier only is feasible if it lies energetically below the entrance channel. However, in many cases that meet this thermochemical requirement rather low efficiencies for the formation of the rearranged products C^+ and D are observed experimentally. This finding was rationalized by the operation of entropic restrictions in the transition structure TS that reduce the probability for rearrangement in comparison to the entropically favorable simple dissociation re-forming the reactants.[108-110] If the rearrangement barrier exceeds the energy available, solely back dissociation occurs unless the energized ion-molecule complex is stabilized by collisional or radiative relaxation.[111] The efficiencies of both processes increase with the lifetime of the initial encounter complex which in turn depends on the effectiveness of internal energy distribution. For systems with a large numbers of internal degrees of freedom, the energy is readily spread over the different vibrational modes such that accumulation in the A^+–B stretching mode leading to dissociation is far more improbable than for smaller systems.[112] Hence, the efficiencies of association reactions are supposed to increase as a function of cluster size.

3   Reactivity of Bare Platinum Clusters Pt_m^+

A systematic investigation into the reactivities of platinum clusters and their potential application as models for heterogeneous catalysts has to start from the very simplest processes, *i.e.*, the reactions of bare platinum clusters Pt_m^+ with small neutral molecules. Only if these basic processes are sufficiently well understood, addressing more complex problems becomes meaningful and feasible. Clearly, the reactions with simple, inexpensive substrates deserve particular interest because activation of such compounds would be of direct relevance to catalysis. The present experiments cover the reactions of Pt_m^+ cluster ions, $m \leq 5$, with the neutral reactants H_2, O_2, CH_4, NH_3, H_2O, CO_2, N_2O, and CH_3NH_2. Although the reactions with CH_4 have been studied before,[43,57] the special importance of the reported activation processes certainly demand an independent verification. With regard to the technically somewhat less relevant compounds N_2O and CH_3NH_2, previous studies on anionic Pt_m^- clusters and atomic Pt^+, respectively,[39,41,61] provoke an extension to the cationic clusters' behavior for the sake of comparison.

A simple classification of the different reactions observed evolves from the general description of ion-molecules reactions given in Section 2.6. For the reactions of platinum clusters Pt_m^+ investigated here, three distinct modes in which the initial collision complex is stabilized are operative.

1.) Association. No atom or molecule is ejected from the initial encounter complex which, instead, is stabilized by photon emission or collisions with the highly diluted neutral acting as third body.

2.) Addition/elimination. The ion-molecule complex is stabilized by eliminating a fragment of the neutral reactant.

3.) Addition/cluster degradation. The initial collision complex is stabilized by loss of a neutral platinum fragment, referred to as cluster degradation.

3.1 Association Reactions

Dihydrogen. For this reaction type, the relaxation mechanisms of the initial ion-molecule encounter complex have been investigated in some detail considering the reaction of platinum clusters with dihydrogen as example. Whereas no reactions are observed for Pt_m^+, $m = 1 - 4$ ($k_{obs} \leq 10^{-13}$ cm^3 s^{-1}, $p(H_2) \approx 10^{-6}$ mbar), association with H_2 takes place in the case of the pentamer, reaction 3.1a.

$$Pt_5^+ + H_2 \quad \rightarrow \quad Pt_5H_2^+ \hspace{3cm} (3.1a)$$

The apparent bimolecular rate constant k_{obs}(3.1a) determined at different pressures $p(H_2)$ clearly shows a linear dependence, which indicates termolecular stabilization of the adduct (Figure 3.1). Extrapolation to $p(H_2) = 0$ gives a positive k value that can be identified with the true bimolecular rate constant. For this, obviously only IR photon emission but no collisional deactivation is operative. Its rather small value, k(3.1a) = $(7 \pm 4) \times 10^{-13}$ cm^3 s^{-1}, which corresponds to a reaction efficiency $\varphi \approx 5 \times 10^{-4}$, implies that most of the initially formed collision complexes dissociate before stabilization by IR emission can occur.

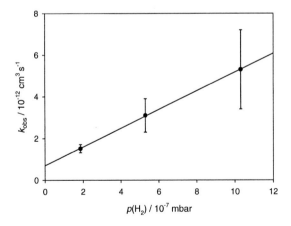

Figure 3.1. Apparent rate constants for the association reaction between Pt_5^+ and H_2 (reaction 3.1a) at three different pressures and linear extrapolation to $p = 0$.

In this context, also a comparison of $k(3.1a)$ with the rate constant of the secondary reaction 3.1b, $k(3.1b)$, is of interest. $Pt_5H_2^+$ reacts approximately twice as efficiently as Pt_5^+, this difference most probably resulting from the higher number of degrees of freedom in the former. As outlined in Section 2.6, such an increase in the system's sum of states is in general supposed to enhance the probability for IR emission and collisional stabilization. The fact that only the largest platinum cluster studied shows a measurable reactivity toward H_2 lends further support to this argument.

$$Pt_5H_2^+ + H_2 \quad \rightarrow \quad Pt_5H_4^+ \tag{3.1b}$$

The rate constant for a third association of H_2 decreases again which indicates that saturation effects begin to come into play. However, the highest uptake of H_2 observed under FT-ICR conditions, namely formation of $Pt_5H_8^+$, still is much lower than that of the clusters $Pt_mH_{5m}^+$ ($m < 25$) found by Kaldor and Cox under flow-tube conditions.[56] Unfortunately, these authors only briefly mention their findings regarding the reactions between Pt_m^+ clusters and H_2, such that no further comparison is possible.

The reaction of Pt_5^+ with D_2 shows similar ratios between the rate constants for the first and the secondary association. Because no pressure extrapolation was attempted, only the apparent rate constant can be compared to that of the corresponding reaction with H_2. Taking into account the different collision rates for H_2 and D_2, a kinetic isotope effect of 1.2 ± 0.3 is obtained for $p \approx 10^{-6}$ mbar. Interestingly, exposure of Pt_5^+ to a mixture of H_2 and D_2 gives rise to $Pt_5H_2^+$, Pt_5HD^+, $Pt_5D_2^+$, along with higher adducts. The occurrence of isotopic scrambling is clear evidence for dissociative chemisorption of H_2 (and D_2) on Pt_5^+, leading from the initial collision complexes to dihydrido-complexes (Scheme 3.1).

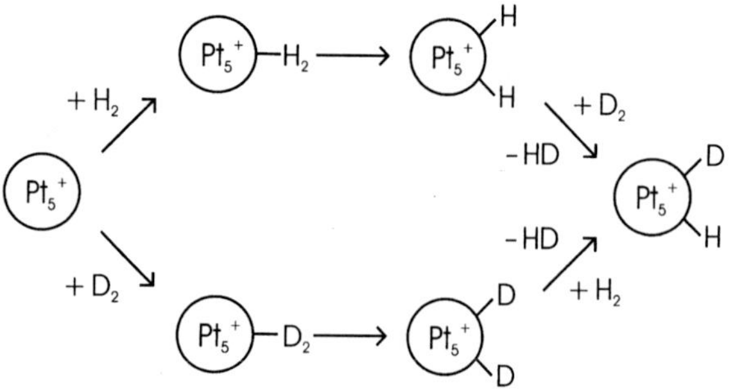

Scheme 3.1.

Ammonia.   Association products are also observed when reacting Pt_m^+ clusters with ammonia, reaction 3.2; a remarkable exception constitutes the platinum dimer whose anomalous reactivity is described in Section 3.2.

$$Pt_m^+ + y\,NH_3 \longrightarrow Pt_mNH_3^+ + (y-1)\,NH_3 \Longrightarrow Pt_mN_yH_{3y}^+ \qquad (3.2)$$

For atomic Pt^+, addition of one NH_3 ligand is so slow under FT-ICR conditions ($k_{obs} \approx 5 \times 10^{-13}$ cm^3 s^{-1}) that the product is hardly detectable, although it has a considerable binding energy, $D_0(Pt^+\!-\!NH_3) = 274 \pm 12$ kJ mol^{-1}.[113] Again, this behavior can be rationalized by the lack of efficient ways to release the excess energy of the collision complex. Enlarging the system by addition of Pt atoms and NH_3 ligands causes significant rate acceleration due to the increased lifetimes of the collision complexes. Pt_5^+, for example, shows an adsorption rate corresponding to $\varphi \approx 0.1$ (Table 3.1). As for H_2, the radiative portion of the observed rate constants might be derived by extrapolation to $p = 0$. For NH_3, however, this method is less practical since the pressure dependence of the association rates appears to be significantly weaker and also the data quality is poorer because of the difficulties in maintaining a constant pressure $p(NH_3)$ during the experiments. Therefore, only apparent rate constants containing termolecular contributions can be reported for reaction 3.2.

For a given cluster size, the association rates pass through a maximum and decrease for higher degrees of ligation (Figure 3.2), thus indicating saturation effects, like in the reactions of Pt_5^+ with H_2. Whereas the trimer Pt_3^+ is observed to add up to four NH_3 molecules according to reaction 3.2, the tetramer and particularly the pentamer do not only undergo simple association but also lose dihydrogen once a certain degree of ligation has been reached, reaction 3.3.

$$Pt_mN_yH_{3y}^+ + NH_3 \quad \rightarrow \quad Pt_mN_{y+1}H_{3y+1}^+ + H_2 \qquad (3.3)$$

In the case of Pt_4^+, the cluster has to add three NH_3 ligands before reaction 3.3 occurs with a fourth molecule (in competition with simple association according to reaction 3.2). For Pt_5^+, this type of reactivity is already observed in the reaction with the second NH_3 molecule, and it becomes the dominating product channel for $y = 2$.[114] Whereas the loss of H_2 inherently implies activation of NH_3 by the platinum cluster, it cannot be judged *a priori* if also in the $Pt_mN_yH_{3y}^+$ species platinum has inserted into an N–H bond. To further investigate this question, Pt_5^+ is exposed to a mixture of NH_3 and D_2. In the resulting products, H/D scrambling occurs only for clusters containing three or more nitrogen atoms, consistent with predominating loss of H_2 at this degree of association. In turn, lacking indications of dissociative chemisorption of either $Pt_5NH_3^+$ or $Pt_5N_2H_6^+$ imply that these contain intact NH_3 molecules.

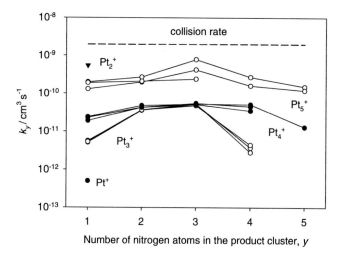

Figure 3.2. Rate constants k_y of the sequential reactions of Pt_m^+ with NH_3.[115] In the case of Pt_2^+, the rate constant given corresponds to dehydrogenation in the primary reaction 3.7a (see Section 3.2). For several reactions, the data of three experiments performed at different pressures are shown.

Water. In the reactions of Pt_m^+ with water, only the pentamer reacts under formation of the corresponding adduct $Pt_5H_2O^+$ at a measurable rate, reaction 3.4 ($k_{obs} = 1.7 \times 10^{-12}$ cm^3 s^{-1}, corresponding to $\varphi = 7 \times 10^{-4}$, $p(H_2O) = 3 \times 10^{-7}$ mbar).

$$Pt_5^+ + H_2O \quad \rightarrow \quad Pt_5H_2O^+ \qquad\qquad (3.4)$$

For the smaller clusters, an upper limit for their rate constants of $k < 2 \times 10^{-13}$ cm^3 s^{-1} is found. Compared with ammonia, the reduced reactivity of water toward cationic platinum clusters can be attributed to its lower basicity because the resulting shallower potential well of the collision complex (DFT predicts $D_0(Pt^+-H_2O) = 180$ kJ mol^{-1} *versus* $D_0(Pt^+-NH_3) =$

310 kJ mol^{-1})[39,116] is supposed to be associated with a decrease in lifetime. The lower number of degrees of freedom for water, compared to ammonia, presumably enhances this effect.

Table 3.1. Apparent rate constants k_{obs} and efficiencies φ for the association reactions between Pt_m^+ and H_2, NH_3, H_2O, and CO_2, respectively.

reaction	m	k_{obs} / cm^3 s^{-1}	$\varphi = k_{obs} / k_c$
$Pt_m^+ + H_2 \rightarrow Pt_mH_2^+$	1 - 4	$\leq 10^{-13}$	$< 10^{-4}$
	5^a	7.0×10^{-13}	5.0×10^{-4}
$Pt_m^+ + NH_3 \rightarrow Pt_mNH_3^+$	1^b	5.0×10^{-13}	3.0×10^{-4}
	3^c	5.4×10^{-12}	3.0×10^{-3}
	4^d	2.3×10^{-11}	1.2×10^{-2}
	5^e	2.0×10^{-10}	0.1
$Pt_m^+ + H_2O \rightarrow Pt_mH_2O^+$	1 - 4	$\leq 2.0 \times 10^{-13}$	$< 10^{-4}$
	5^f	1.7×10^{-12}	7.0×10^{-4}
$Pt_m^+ + CO_2 \rightarrow Pt_mCO_2^+$	1 - 5	$\leq 5.0 \times 10^{-13}$	$< 10^{-3}$

^{a} Bimolecular rate constant, see text. ^{b} Taken from ref. [39]. ^{c} No significant differences in k_{obs} for $p = 7 \times 10^{-8}$, 3×10^{-7}, and 8×10^{-7} mbar. Average value given for k_{obs}. ^{d} No significant differences in k_{obs} for $p = 8 \times 10^{-8}$, 2×10^{-7}, and 5×10^{-7} mbar. Average value given for k_{obs}. ^{e} No significant differences in k_{obs} for $p = 8 \times 10^{-9}$, 2×10^{-8}, and 5×10^{-8} mbar. Average value given for k_{obs}. $^f p = 3 \times 10^{-7}$ mbar.

Carbon dioxide.   Finally, in the reactions of Pt_m^+ with carbon dioxide, association products are not detected even for the pentamer. However, for Pt_4^+ und Pt_5^+, degradation according to reaction 3.5 cannot be excluded rigorously as this type of reactivity might be obscured by the inevitable presence of traces of O_2 readily yielding the same ionic products (see Section 3.3). Nevertheless, upper limits for the efficiency of reaction 3.5 can be derived as $\varphi \leq 5 \times 10^{-4}$ and $\leq 5 \times 10^{-3}$ for the tetramer and pentamer, respectively.

$$Pt_m^+ + CO_2 \quad \rightarrow \quad Pt_{m-1}^+ + [Pt,C,O_2] \tag{3.5}$$

3.2   Addition/Elimination Reactions

Methane.   The prototype of this reaction in the case of platinum clusters is dehydrogenation of methane according to reaction 3.6. The present findings are in good agreement with the results of Bondybey and co-workers (Table 3.2).[57] In particular, the previously reported anomalously low reactivity of the tetramer is fully confirmed. Possible reasons for this distinct behavior are addressed in Chapter 4 whereas the question of consecutive reactions is treated in Chapter 5.

$$Pt_m^+ + CH_4 \quad \rightarrow \quad Pt_mCH_2^+ + H_2 \tag{3.6}$$

Ammonia.   Similarly to reaction 3.6, the platinum dimer Pt_2^+ induces dehydrogenation of NH_3, reaction 3.7a. The resulting ionic product, Pt_2NH^+, then transfers its proton to a further NH_3 molelcule, reaction 3.7b.

$$Pt_2^+ + NH_3 \quad \rightarrow \quad Pt_2NH^+ + H_2 \tag{3.7a}$$
$$Pt_2NH^+ + NH_3 \quad \rightarrow \quad NH_4^+ + Pt_2N \tag{3.7b}$$

As all other Pt_m^+ cluster ions investigated undergo mere association, the question arises why just the dimer behaves differently. It is interesting to note that a similar case has been reported for rhodium clusters Rh_m^+. Whereas neither the monomer nor the trimer or larger clusters Rh_m^+ react with methane, the dimer efficiently yields the corresponding carbene $Rh_2CH_2^+$ in analogy to reaction 3.6.[117] A rationale for this strongly size-specific reactivity could be that a metal dimer, thanks to its lower degree of coordination, possesses a higher intrinsic potential for bond activation than the larger clusters. Compared with the monomer, the dinuclear ion presumably gains more energy from the initial interaction with the substrate, thus helping to surmount the critical reaction barriers and suffice the thermochemical requirements associated with dehydrogenation. Moreover, the presence of two metal atoms that can form bonds towards the divalent dehydrogenation fragment, *i.e.*, CH_2 or NH, respectively, is expected to enhance stabilization of the product and thereby further assist in its generation. In the case of the reaction between Pt_2^+ and NH_3, the product may be thought of as an imine. The increase in acidity well-known to be associated with the change from sp^3 to sp^2 hybridization indeed agrees with the consecutive protolysis, which is observed for Pt_2NH^+, reaction 3.7b, but not for $Pt_mNH_3^+$ ($m = 1, 3 - 5$).

Table 3.2. Bimolecular rate constants k and efficiencies φ for the addition/elimination reactions of Pt$_m^+$ with CH$_4$, NH$_3$, CH$_3$NH$_2$, and N$_2$O, respectively.

reaction	m	k / cm^3 s^{-1}	φ
Pt$_m^+$ + CH$_4$ $\rightarrow$ Pt$_m$CH$_2^+$ + H$_2$	1	$5.0 \times 10^{-10\,a}$	0.51
	2	$8.2 \times 10^{-10\,b}$	0.85
	3	$6.0 \times 10^{-10\,c}$	0.63
	4	$1.5 \times 10^{-11\,d}$	0.02
	5	$8.8 \times 10^{-10\,e}$	0.93
Pt$_m^+$ + NH$_3$ $\rightarrow$ Pt$_m$NH$^+$ + H$_2$	2	5.4×10^{-10}	0.27
Pt$_m$NH$^+$+ NH$_3$ $\rightarrow$ NH$_4^+$ + Pt$_m$N	2	6.6×10^{-10}	0.33
Pt$_m^+$ + CH$_3$NH$_2$ $\rightarrow$ [Pt$_m$,C,H,N]$^+$ + 2 H$_2$	1	9.5×10^{-11}	0.06
	2	7.5×10^{-10}	0.50
	3	1.4×10^{-9}	0.92
	4	1.0×10^{-9}	0.68
	5	1.1×10^{-9}	0.72
Pt$_m^+$ + CH$_3$NH$_2$ $\rightarrow$ [Pt$_m$,C,H$_3$,N]$^+$ + H$_2$	1	5.2×10^{-11}	0.03
	2	5.0×10^{-11}	0.03
Pt$_m^+$ + CH$_3$NH$_2$ $\rightarrow$ CH$_2$NH$_2^+$ + Pt$_m$H	1	7.1×10^{-10}	0.46
	2	2.0×10^{-10}	0.13
Pt$_m^+$ + N$_2$O $\rightarrow$ Pt$_m$O$^+$ + N$_2$	1	$7.0 \times 10^{-11\,f}$	0.10
	2	$\leq 2.0 \times 10^{-13}$	$< 3.0 \times 10^{-4}$
	3	5.5×10^{-12}	1.3×10^{-3}
	4	$\leq 3.0 \times 10^{-13}$	$< 5.0 \times 10^{-4}$
	5	1.6×10^{-11}	0.02
Pt$_m$O$^+$ + N$_2$O $\rightarrow$ Pt$_m$O$_2^+$ + N$_2$	1	$1.9 \times 10^{-10\,f}$	0.27
	3	1.2×10^{-11}	0.02
	5	5.8×10^{-11}	0.09

[a] Compare with $k = 3.9 \pm 1.0$ (ref. [33]), 5.9 ± 2.5 (ref. [35], revised according to ref. [37]), 4.6 ± 0.3 (ref.[57]), and 7.6 ± 1.7 ([37], for $E_{kin} = 0.05$ eV), all values in 10^{-10} cm^3 s^{-1}. [b] Compare with $k = (7.0 \pm 0.6) \times 10^{-10}$ cm^3 s^{-1} (ref. [57]). [c] Compare with $k = (8.7 \pm 0.7) \times 10^{-10}$ cm^3 s^{-1} (ref. [57]). [d] Compare with $k = (2.7 \pm 0.7) \times 10^{-11}$ cm^3 s^{-1} (ref. [57]). [e] Compare with $k = (1.1 \pm 0.2) \times 10^{-9}$ cm^3 s^{-1} (ref. [57]). [f] Taken from (ref. [41]).

Methylamine.   Compared to CH$_4$, the presence of an amino group in methylamine is expected to increase the substrate's reactivity and to further facilitate dehydrogenation. Indeed, the predominant reaction of Pt$_m^+$ clusters with CH$_3$NH$_2$ leads to double dehydrogenation according to reaction 3.8a. For atomic Pt$^+$, single dehydrogenation takes place as well, reaction 3.8b, although the main reaction (80 % branching ratio, b.r.) observed for this ion corresponds to charge transfer from the metal to the organic fragment by formal hydride migration, reaction 3.8c. In the case of the dimer ion Pt_2^+, reactions 3.8b and 3.8c still occur but are outweighed by the double dehydrogenation 3.8a (5 *vs* 20 *vs* 75 % b.r.); for the larger clusters, only the latter is observed. Apparently, hydride transfer inversely correlates with the ionization energies (*IEs*) of the platinum clusters (*IE*(Pt) = 9.0 eV,[118] *IE*(Pt$_2$) = 8.68 ± 0.02 eV;[119] for the larger clusters, accommodation of a positive charge is assumed to be even easier).

$$Pt_m^+ + CH_3NH_2 \quad \rightarrow \quad [Pt_m,C,H,N]^+ + 2\,H_2 \qquad (3.8a)$$

$$\rightarrow \quad [Pt_m,C,H_3,N]^+ + H_2 \qquad (3.8b)$$

$$\rightarrow \quad CH_2NH_2^+ + Pt_mH \qquad (3.8c)$$

In the case of mononuclear Pt^+, previous labeling experiments imply occurrence of a 1,1-elimination leading to an aminocarbene structure $PtC(H)NH_2^+$ in reaction 3.8b rather than a 1,2-dehydrogenation.[120] Thus, Pt^+ selectively attacks the methyl group in CH_3NH_2 which is consistent with its high reactivity toward CH_4 and its failure to dehydrogenate NH_3. Double dehydrogenation results in an isonitrile complex $Pt(CNH)^+$.[120] For the Pt_m^+ cluster ions, the predominance of double dehydrogenation points to low barriers associated with the second H_2 elimination. CID of the product ions $[Pt_m,C,H,N]^+$ solely affords simple loss of HCN or HNC, as explicitly studied for $m = 1$,[120] 4 and 5, and thus is not particularly indicative with respect to the structural question. In the case of the Pt_2^+ dimer, exposure to labeled CD_3NH_2[121] yields $[Pt_2,C,H,N]^+$ after double dehydrogenation (along with $CD_2NH_2^+$ as second significant ionic product), which clearly suggests formation of an isonitrile species, much like for $m = 1$. Assuming analogous isonitrile structures for the heavier clusters as well, the second dehydrogenation step occurring fast for the clusters corresponds to a 1,2-elimination from aminocarbene complexes $Pt_mC(H)NH_2^+$. The presence of a larger metal core in the case of the clusters might enable a secondary interaction with the amino group which would be likely to assist in 1,2-dehydrogenation and can thus account for the experimental findings.

The apparently low barriers associated with the dehydrogenation reactions are also mirrored in the high reaction rates observed for these processes (Table 3.2). Essentially, unity efficiency is approached for all cluster sizes (in the case of Pt^+, the competing hydride transfer provides the largest contribution to the overall rate constant). This situation contrasts with the behavior of Pt_m^+ clusters toward CH_4 where the tetramer shows an exceptional low reactivity (see above). Obviously, the enhanced tendency of CH_3NH_2 towards H_2 elimination largely levels distinctions in reactivity between the different cluster sizes. Moreover, the Pt_m^+ ions do not stop to react with CH_3NH_2 after dehydrogenation of the first molecule but continue to undergo further additions. For comparable pressures and reaction times, the maximal number of added molecules x increases as a function of cluster size: $x = 3$ for $m = 1, 2$, $x = 4$ for $m = 3$, $x = 5$ for $m = 4$, and finally $x = 7$ for $m = 5$. Whereas the first CH_3NH_2 molecules added are predominantly dehydrogenated twice, single H_2 elimination prevails for the highly ligated clusters and leads to $[Pt_m,C_x,H_{x+2},N_x]^+$ and $[Pt_m,C_x,H_{x+4},N_x]^+$ species, reactions 3.9a and 3.9b. The relatively large HNC (or HCN) uptake observed for the clusters is consistent with formation of dative rather than covalent bonds, as expected for this closed-shell ligand.

$$Pt_m^+ + x\,CH_3NH_2 \quad \rightarrow \quad [Pt_m,C_x,H_{x+2},N_x]^+ + (2x-1)\,H_2 \qquad (3.9a)$$

$$\rightarrow \quad [Pt_m,C_x,H_{x+4},N_x]^+ + (2x-2)\,H_2 \qquad (3.9b)$$

For kinetic analyses of the overall reaction schemes, simplified models that do not distinguish between double and single dehydrogenation are applied. Regarding the only moderate data quality,[122] a more sophisticated approach is not indicated. Also, the analyses do not reveal any unexpected findings but yield rate constants for the consecutive reactions corresponding to efficiencies $\varphi \approx 0.5 - 1$; only for the final steps, slightly lower values are derived which point to saturation effects. Hence, high reactivities without size-specific effects can be inferred for the ligated platinum clusters as well. As a result of side reactions, consecutive formation of $CH_3NH_3^+$ is observed to a minor extent. Such protolysis reactions do not come unexpected, given the basicity of CH_3NH_2 and the presumably appreciable acidity of CNH of HCN species bound to cationic metal centers. In the cases of Pt^+ and Pt_2^+, additional amounts of $CH_3NH_3^+$ arise from proton transfer from the primary product $CH_2NH_2^+$. This process simply reflects the higher gas-phase basicity of CH_3NH_2 compared to CH_2NH (864.5 *versus* 818.8 kJ mol^{-1}).[123]

Nitric oxide.   As expected, N_2O reacts with Pt_m^+ by oxygen-atom transfer, reaction 3.10.[124]

$$Pt_m^+ + y\,N_2O \longrightarrow Pt_mO^+ + (y-1)\,N_2O + N_2 \rightleftharpoons Pt_mO_y^+ + y\,N_2 \qquad (3.10)$$

With respect to thermodynamics, reaction 3.10 is highly favorable because both a relatively strong metal-oxide bond is formed and the particularly stable N_2 molecule is released. However, the rate constants actually observed for the primary reaction are rather low for the different cluster sizes examined (Table 3.2). For the dimer and tetramer, the occurrence of reaction 3.10 cannot be established at all ($\varphi < 5 \times 10^{-4}$), and even the most reactive species, *i.e.*, atomic Pt^+, does not exhibit an efficiency higher than $\varphi = 0.1$.[41] Systematic studies by Armentrout *et al.*[125] and Schwarz and co-workers[126] suggest that effectiveness of barriers is quite a general phenomenon for O-atom transfer from N_2O to transition-metal cations M^+. The qualitative explanation put forward by these authors notes that the ground states of cationic oxides MO^+ correlate with atomic $O(^3P)$ whereas adiabatic oxygen release from N_2O yields $O(^1D)$, such that the overall reactions leading to MO^+ in their ground states are formally spin-forbidden.[125,126] Although the potential-energy surfaces of the joined systems, $M^+ + N_2O$, are certainly much more complicated,[127] particularly if not only single metal atoms but clusters are involved, there is experimental evidence that barriers indeed are a common feature of reactions between M_m^+ and molecules XYZ being isoelectronic with N_2O.[128,129] For the reactions of N_2O and anionic transition-metal clusters, the situation is complex too. Whereas Pd_m^- clusters induce O-atom transfers with high efficiencies, the analogous reactions are strongly size-specific in the case of Pt_m^-. Hintz and Ervin found the trimer and the hexamer to be more than one order of magnitude less reactive than the other clusters investigated ($m = 3 - 7$).[61] Probably, a detailed understanding of O-transfer in reaction 3.10 and related problems has to await substantial progress in the theoretical treatment of platinum clusters.

Secondary O-atom transfers are observed up to $y = 2$ for $m = 1$ and 5, respectively, and $y = 3$ for $m = 3$, reaction 3.10. In the case of the trimer and pentamer, these numbers should not necessarily be considered as saturation level, but they nevertheless suggest that any possible further consecutive reactions are rather inefficient. For the mononuclear system, previous labeling experiments gave strong evidence for an inserted structure $OPtO^+$, thus explaining the absence of a third O-transfer which would result in an unreasonably high oxidation state of platinum.[41] Remarkably, in all cases the secondary O-transfer occurs significantly more efficiently than the primary one, indicating that the barriers associated with the former are less pronounced.

3.3   Addition/Cluster Degradation Reactions

This reaction becomes operative if the interaction between the cluster and the substrate is highly exothermic and expulsion of a cluster fragment is the energetically least demanding exit channel. Such a reaction is observed to take place between platinum clusters Pt$_m$$^+$ and dioxygen, reaction 3.11a.

$$Pt_m^+ + O_2 \rightarrow Pt_{m-1}^+ + PtO_2 \tag{3.11a}$$

Although the neutral product PtO$_2$ cannot be detected, its generation rather than that of the separated atoms is unambiguous because the latter reaction would imply the occurrence of an endothermic process. Interestingly, the rate constants show a marked dependence on cluster size (Table 3.3). Whereas Pt$_m$$^+$ clusters, $m = 2$, 4, and 5, react fast (for the pentamer, unity efficiency is reached), the trimer reacts more than 100 times more slowly. In the absence of considerable barriers, the drastic decrease in efficiency would reflect a particular thermochemical stability of Pt$_3$$^+$, speculatively because of distinct electronic or geometric features of this cluster size. Notably, the tetramer does not behave anomalously in its reaction with O$_2$, unlike to that with CH$_4$ (see Section 3.2). Hence, size-specific reactivity appears to depend crucially on the particular substrate in the case of platinum clusters, such that general conclusions based on the reactions with a single substrate might be misleading.

Another interesting aspect of reaction 3.11a is related with the oxygen distribution in the products. Taking into account that the reaction passes through the transient collision complex Pt$_m$O$_2$$^+$, the final elimination step can be considered as a disproportionation. This behavior resembles chemistry of platinum in solution where Pt^{IV} is well-known to be generally more stable than Pt^{II}.[130] Other product channels are only observed in the case of the trimer, reactions 3.11b and 3.11c. Note, however, that these are even less efficient than reaction 3.11a for $m = 3$.

$$Pt_3^+ + O_2 \rightarrow Pt_2O^+ + PtO \tag{3.11b}$$

$$Pt_3^+ + O_2 \rightarrow Pt_2O_2^+ + Pt \tag{3.11c}$$

For atomic Pt$^+$, simple addition of O$_2$ has been observed under flow-tube conditions.[131] Despite the higher probability of termolecular stabilization in these experiments, the association efficiency is as low as $\varphi = 3 \times 10^{-4}$. Moreover, chemisorption of O$_2$ as well as fragmentation has been observed for the anionic clusters Pt$_m$$^-$ in flow-tube experiments.[61]

Fragmentation predominates for the smallest cluster sizes studied, namely the trimer and the tetramer. In the case of the former, the overall reaction rate is much lower than that for the other clusters, similarly to the situation for its cationic homologue. However, without having further data for comparison, the origin of this finding remains unclear.

Table 3.3. Bimolecular rate constants k and efficiencies φ for the degradation reactions of Pt_m^+ exposed to O_2.

reaction	m	$k\,/\,\mathrm{cm^3\,s^{-1}}$	φ
$Pt_m^+ + O_2 \rightarrow Pt_{m-1}^+ + PtO_2$	2	1.3×10^{-10}	0.23
	3	7.6×10^{-13}	1.4×10^{-3}
	4	1.4×10^{-10}	0.25
	5	5.8×10^{-10}	1.1
$Pt_m^+ + O_2 \rightarrow Pt_{m-1}O^+ + PtO$	3	7.0×10^{-14}	1.0×10^{-4}
$Pt_m^+ + O_2 \rightarrow Pt_{m-1}O_2^+ + Pt$	3	5.0×10^{-13}	9.0×10^{-4}

3.4   Implications with Respect to Heterogeneous Catalysis

At first, the present results can be compared with the reactions observed for atomic Pt^+ as a much simpler gas-phase model for extended platinum surfaces. Neglecting degradation reactions, which are naturally only possible for the clusters, one is left with association and addition/elimination reactions. The efficiencies of the former strongly increase when going from the monomer to the larger clusters because of lifetime effects. As their importance is quite a general phenomenon in cluster studies performed in the highly-diluted gas phase, they do not point to a peculiarity inherent in platinum. However, because such simple association reactions are considered as important elementary steps in heterogeneous catalysis, their occurrence in the case of the clusters opens up interesting prospects, as demonstrated for the reactions with H_2 and NH_3. In this respect, the clusters are obviously superior to mononuclear ions with regard to their suitability as model systems. In contrast, the addition/elimination reactions with CH_4 and N_2O, respectively, are similar in the case of atomic Pt^+ and Pt_m^+ clusters which indicates that their occurrence does not significantly depend on the ion's charge density. The same conclusion can be inferred from the reactions of these substrates with anionic Pt_m^- clusters;[57,61] although the efficiencies differ, the products formed are analogous to those generated from Pt$_m^+$. A stronger effect of the ion's charge density on the outcome of the reaction is observed in the case of CH$_3$NH$_2$, however, where hydride transfer only occurs for Pt$^+$ and Pt$_2^+$.

Certainly, a comparison between the reactions studied here and those taking place at the surface of bulk platinum is even more warranted for an assessment of the clusters' potential as model systems. In the case of H$_2$, the dissociative chemisorption observed to occur on Pt$_5^+$ indeed reflects the analogous process known for the condensed phase.[132] Moreover, activation of CH$_4$ as found for Pt$_m^+$ also occurs on certain heterogeneous platinum catalysts.[133] In the DEGUSSA and Andrussov process, this distinct reactivity of platinum is used for the synthesis of hydrogen cyanide from CH$_4$ and NH$_3$. However, the very limited mechanistic insight into these catalytic reactions does not permit a more thorough comparison with the results from the gas-phase studies.[17-22]

For the reactions between platinum and NH$_3$, quite a complicated picture emerges. Whereas most of the clusters investigated simply add NH$_3$ in the primary reaction, a remarkable exception is the dimer Pt$_2^+$ which readily effects dehydrogenation. Notably, platinum crystal-faces predominately adsorb molecular NH$_3$, but also induce dissociative chemisorption whose efficiency strongly depends on the structure of the Pt surface.[134,135] However, dissociation of NH$_3$ does not stop at the stage of the imine but continues towards total decomposition of the substrate, eventually followed by desorption of N$_2$ and H$_2$, a process apparently not operative in the case of Pt$_2^+$. Interestingly, at $T > 1200$ K, free NH radicals were observed to be formed upon exposure of pure NH$_3$ to clean Pt surfaces which implies a dehydrogenation of the substrate in a previous step.[136] Considering the presumably high interaction energy between Pt$_2^+$ and NH$_3$ (compare D_0(Pt$^+$–NH$_3$) = 274 ± 12 kJ mol^{-1})[113] which is distributed over a limited number of internal degrees of freedom, a high effective temperature results for the transient complex, thus rationalizing the similar reactivities. In the case of Pt$_4^+$ and Pt$_5^+$, activation of NH$_3$ only occurs after adsorption of several substrate molecules. Translated into condensed phase, this would correspond with NH$_3$ activation beginning at higher coverage only; a phenomenon so far unknown in surface studies.

In the case of CH$_3$NH$_2$, platinum clusters readily induce double dehydrogenation. The same behavior was observed for polycrystalline platinum contacts at elevated temperatures.[137] Whereas the gas-phase experiments do not distinguish between

$Pt_m(NCH)^+$ and $Pt_m(CNH)^+$ structures, formation of HCN as the thermodynamically favored isomer occurs under heterogeneous conditions. In both systems, no indications of C–N bond breaking are found whereas such processes play an important role for the decomposition of CH_3NH_2 on other transition-metal surfaces.[138]

Concerning the reactivity of platinum surfaces towards N_2O, earlier studies at low temperatures found adsorption to be inefficient and accompanied by a minor extent of decomposition only.[139,140] In contrast, application of higher N_2O pressures at 363 K leads to formation of a complete oxygen monolayer on the Pt surface.[141] As O-atom transfer also occurs in the reactions between N_2O and $Pt_m{}^+$, another analogy connecting Pt clusters and surface chemistry evolves. The fact that N_2O decomposition on bulk Pt does not take place at low temperatures points to the operation of activation barriers which are also apparent in the corresponding reactions of $Pt_m{}^+$.

Finally, cluster fragmentation observed in the reactions with O_2 implies that the interaction between $Pt_m{}^+$ and O_2 releases enough energy to degrade the metal cluster, thus only being compatible with an inserted structure of the intermediate $Pt_mO_2{}^+$ complex. Accordingly, a recent DFT study predicted $Pt_3{}^+$ to cleave the O–O bond in O_2 on energetic reasons.[142] For Pt crystal faces, dissociative chemisorption of O_2 is well-known to predominate at $T > 150$ K.[143-145] In marked contrast to the situation encountered in gas-phase studies of Pt clusters, however, the reaction energy released can be dissipated by the solid, thereby preventing surface degradation. Interestingly, at $T > 1000$ K, the Ostwald catalysts used to oxidize NH_3 by O_2 show a leaching of Pt which indeed was rationalized by generation of PtO_2 (g).[146,147]

As a preliminary conclusion, small $Pt_m{}^+$ clusters are found to exhibit a reactivity much resembling that of platinum surfaces. Hence, investigations of the reactions of these clusters appear as a reasonable approach to model the far more complex processes in heterogeneous catalysis at least in the case of platinum. Particularly, $Pt_m{}^+$ clusters mimic the bulk surface in several aspects more closely than mononuclear Pt^+ ions. However, one must also be aware of the specificities of certain $Pt_m{}^+$ clusters. In the reactions with CH_4, NH_3, and O_2, anomalous reactivities are observed for a single cluster size in each case, thus indicating size-specific electronic or geometric effects which have been inferred for other transition-metal clusters as well. For platinum, however, the situation seems to be particularly complicated as the exceptional cluster size differs in each reaction. Further experimental and especially theoretical work is necessary to elucidate the origin of this behavior. For the time being, any conclusions about structural peculiarities of a certain cluster size derived from anomalies in reactivity should be viewed with caution.

Having demonstrated the basic validity of the present approach, more complex problems can be tackled. A principal objective concerns the simultaneous usage of a second substrate which would offer manifold opportunities. Among these, functionalization of methane appears most attractive because of its tremendous economic relevance as outlined in Chapter 1. In the case of mononuclear Pt^+, the primary product from CH_4 dehydrogenation, *i.e.*, $PtCH_2^+$, can indeed be coupled with ammonia and other nucleophilic substrates.[39,40,148] The question if analogous processes also apply to the clusters ions $Pt_mCH_2^+$ will form one of the main issues in the further course of the present thesis. However, prior to this undertaking, a more detailed investigation of the Pt_m^+/CH_4 system clearly is indicated.

4 Experimental Probes for the Pt_m^+/CH_4 Potential-Energy Surfaces

Irikura's and Beauchamp's observation of methane activation by platinum and other third-row transition-metal ions reported in 1991 marks the beginning of extensive research devoted to these intriguing reactions.[32,33] Whereas, with the exception of Zr^+,[149,150] none of the first and second-row transition-metal mono-cations succeeds in the activation of methane at thermal energies, their third-row congeners Ta^+, W^+, Os^+, Ir^+, and Pt^+ readily accomplish dehydrogenation of CH_4, reaction 4.1.[33]

$$M^+ + CH_4 \quad \rightarrow \quad MCH_2^+ + H_2 \qquad\qquad (4.1)$$

There is general agreement that relativistic effects account for the distinct behavior of these $5d$ metals.[151] Essentially, relativity lowers the $6s$ and lifts the $5d$ orbitals in energy, thus favoring sd hybridization. As a result, the promotion energies for achieving $d^{n-1}s^1$ configurations are relatively small such that a strong binding to the methylene fragment evolves.[33] Note that the dehydrogenation 4.1 only becomes exothermic for bond-dissociation energies $D_{298}(M^+-CH_2) > 465 \text{ kJ mol}^{-1}$.[123]

In the case of platinum, additional experimental and theoretical efforts have been made to elucidate the mechanism of reaction 4.1 and to map the system's potential-energy surface (PES).[35,36,116] A recent work by Zhang et al. summarizes these studies and brings them to a preliminary end.[37] Briefly, the reaction between Pt^+ and CH_4 is found to initially lead to a $Pt^+(CH_4)$ ion-molecule complex which is separated by an energetically very low barrier from the inserted structure $H-Pt^+-CH_3$. The latter corresponds to the global minimum of the PES that lies $171 \pm 8 \text{ kJ mol}^{-1}$ below the entrance channel. Rearrangement via a dihydrido species $(H)_2PtCH_2^+$ yields $(H_2)PtCH_2^+$ which gives the final product $PtCH_2^+$ after H_2 elimination. All barriers are lower in energy than the exit channel which, in turn, lies below the $Pt^+ + CH_4$ asymptote by 9 kJ mol^{-1}, thus rendering the overall reaction slightly exothermic.[37]

With respect to the reactions of the homologous clusters, their PES still await characterization. In comparison with the mononuclear system, the increased complexity of the clusters constitutes a tremendous challenge particularly for theoretical methods. It might in fact be doubted if state-of-the-art quantum-chemical techniques could reach at all a reliable description of these systems. Certainly, a definite answer to this question remains difficult as long as experimental benchmarks are not available. On the side of experiment, guided-ion beam techniques have been proven highly valuable for obtaining thermochemical quantities.[152] Because these experiments rely on quadrupole mass-detection, however, the latter's limited mass resolution hinders the distinction between Pt$_m$CH$_x$$^+$ or even Pt$_m$CD$_x$$^+$ species. FT-ICR mass spectrometry is superior in this regard but does not achieve comparable thermochemical accuracy on the other hand. Nevertheless, it should allow studying the energy-dependence of CID of Pt$_m$CH$_2$$^+$, $m \leq 5$, at a semi-quantitative level which is already expected to be instructive in the present context. Prior to this approach, the kinetic isotope effects (KIEs) associated with reaction 4.1 and its deuterium-labeled analogue as well as the reverse reactions are investigated. These experiments promise to shed further light on the systems' PESs.

4.1   Reactions of Pt$_m$$^+$ with CH$_2$D$_2$

The great relevance of KIEs to mechanistic investigations arises from their sensitivity to the underlying PES.[153] Within the Born-Oppenheimer approximation, the masses of the nuclei do not influence the PES itself, whereas they directly affect the vibrational frequencies ν according to

$$\nu = \frac{1}{2\pi}\sqrt{\frac{D}{\mu}} \qquad\qquad (4.2)$$

where D is the force constant and μ the reduced mass. For large relative mass differences such as in the case of H versus D, considerably different zero-point energies $E_0 = \frac{1}{2}$ hν may therefore significantly influence the effective barrier heights associated with cleavage of the corresponding bond. The lower mass of H compared to D yields a larger E_0 such that the effective barrier for dissociation is reduced and the reaction can proceed more readily. If breakage of the very bond forms the rate-determining step of the overall process, a primary KIE $\equiv k_H/k_D \gg 1$ is observed. At ambient temperatures, typically KIE ≈ 7 is found for the rate-limiting dissociation of a C–H bond.[153]

The rate constants k_H and k_D can be compared most easily if the corresponding reactions directly compete within the same system. The observed product distributions then simply reflect the intramolecular KIEs. With respect to the present problem, use of CH_2D_2 is particularly helpful. Reaction with Pt_m^+ gives rise to three different products, reactions 4.3a - c. Note that the statistical weights of the individual product channels have to be acknowledged (1:4:1 for reactions 4.3a - c) for the derivation of the KIEs.[154]

$$Pt_m^+ + CH_2D_2 \quad \rightarrow \quad Pt_mCH_2^+ + D_2 \quad\quad (4.3a)$$

$$\rightarrow \quad Pt_mCHD^+ + HD \quad\quad (4.3b)$$

$$\rightarrow \quad Pt_mCD_2^+ + H_2 \quad\quad (4.3c)$$

The observed KIEs do not strongly depend on cluster size m (Table 4.1). Moreover, they are all relatively small and, thus, suggest that hydrogen migrations are not involved in the rate-determining steps. Similarly, the PES proposed by Zhang *et al.* for the system Pt^+/CH_4 assigns the highest barrier to elimination of H_2 from a $(H_2)PtCH_2^+$ structure;[37] an analogous situation may be assumed for the clusters as already proposed by Achatz *et al.*[57]

Table 4.1. Intramolecular KIEs[a] of the reactions of Pt_m^+ with CH_2D_2, reactions 4.3a - c.

	$m = 1$	2	3	4	5
KIE(4.3b/4.3a)	1.7 ± 0.2^b	1.2 ± 0.1	1.1 ± 0.2	1.3 ± 0.1	1.0 ± 0.1
KIE(4.3c/4.3b)	1.7 ± 0.1^b	1.2 ± 0.1	1.5 ± 0.1	1.3 ± 0.2	1.4 ± 0.2

[a] Corrected for statistical effects, see text. [b] Compare with KIE(4.3b/4.3a) = KIE(4.3c/4.3b) = 1.6 reported in [34].

4.2 Reactions of $Pt_mCH_2^+$ with H_2

Further information can be obtained by accessing the Pt_m^+/CH_4 system from an alternative entrance channel. First, this approach should reveal if the reverse process of methane dehydrogenation, reaction 4.4, is feasible.

$$Pt_mCH_2^+ + H_2 \quad \rightarrow \quad Pt_m^+ + CH_4 \quad\quad (4.4)$$

As already established previously, reaction 4.4 occurs with low efficiency for mononuclear platinum which is consistent with the only very small endothermicity $\Delta_r H° = 9$ kJ mol^{-1} associated with this process.[35]

Reaction 4.4 is not observed for cluster sizes $m = 2$, 3, and 5 which might indicate significant endothermicities in these cases. In line with this interpretation, the reverse process, *i.e.*, dehydrogenation 4.3, takes place with somewhat higher efficiencies for these clusters compared to Pt$^+$ (Table 3.2). In marked contrast, the tetramer Pt$_4^+$ displays an anomalously low reactivity toward CH$_4$. Interestingly, for this cluster size reaction 4.4 occurs with a considerable efficiency. From the ratio of rate constants, the corresponding equilibrium constant, $K = k(4.3)/k(4.4) = 0.9 \pm 0.4$ and, thereby, the free reaction enthalpy, $\Delta_r G°(298$ K$) = 0 \pm 2$ kJ mol^{-1} can be derived.[155] Note that the $\Delta_r G°$ value determined does not reflect thermochemistry only, but that also entropy changes play a role, as one vibrational degree of freedom is transformed into a rotational one in the course of the reaction. Given the much lower rotational excitation energies, this change is supposed to increase the system's sum of states and, thus, to favor reaction 4.4 compared to 4.3.

Apparently, the lack of a thermodynamic driving force is an important factor contributing to the low efficiency of methane dehydrogenation 4.3 in the case of the tetramer. This conclusion is also consistent with the finding of Hanmura *et al.* that for the tetramer reaction 4.3 is not anomalously slow at 0.15 eV collision energy, as the provision of kinetic energy is supposed to compensate for the insufficient thermodynamic impulsion of the reaction.[31] However, the origin of the distinct thermodynamics found for Pt$_4^+$/CH$_4$ remains unclear. Preliminary DFT calculations by Achatz *et al.* predict a tetrahedral structure for Pt$_4^+$ but this result alone does not provide an explanation either.[57]

4.3   H/D Exchange Processes

Besides formation of CH$_4$, the reaction of H$_2$ with Pt$_m$CH$_2^+$ should also lead to degenerate hydrogen exchange. Usage of D$_2$ enables monitoring of the exchange processes, reactions 4.5a - c.

$$\text{Pt}_m\text{CH}_2^+ + \text{D}_2 \quad \rightarrow \quad \text{Pt}_m\text{CHD}^+ + \text{HD} \qquad (4.5a)$$

$$\text{Pt}_m\text{CHD}^+ + \text{D}_2 \quad \rightarrow \quad \text{Pt}_m\text{CD}_2^+ + \text{HD} \qquad (4.5b)$$

$$Pt_mCH_2^+ + D_2 \quad \rightarrow \quad Pt_mCD_2^+ + H_2 \tag{4.5c}$$

Reactions 4.5a and 4.5b constitute a stepwise exchange, whereas reaction 4.5c corresponds to direct substitution of H_2 for D_2. As these processes involve either identical reactant or product ions, the determination of separate rate constants requires modeling. Prior to such a quantitative treatment, a qualitative discussion of the experimental findings is indicated.

The exchange processes observed for $PtCH_2^+$ and $Pt_2CH_2^+$ show close similarities (Figures 4.1 and 4.2).[156] The occurrence of Pt_mCHD^+ species in both cases is clear evidence for stepwise substitution 4.5a whereas the operation of direct exchange 4.5c cannot be inferred without modeling. However, the Pt_mCHD^+ ions do not reach relative abundances > 30 % which points to significant efficiencies of the consecutive reactions 4.5b. In contrast, the analogous reaction of $Pt_3CH_2^+$ leads to a maximum relative abundance of Pt_3CHD^+ as high as 70 % (Figure 4.3). Obviously, the trinuclear cluster behaves quite distinctly from its smaller homologues. Given the similar KIEs observed for these cluster sizes in reactions 4.3a - c, this difference is rather surprising. In the case of $Pt_4CH_2^+$, the situation is severely complicated by the occurrence of side reactions. First, reaction 4.4 yielding Pt_4^+ has a considerable efficiency for this cluster size. Moreover, $Pt_4CH_2^+$ reacts fast with traces of background water according to reaction 4.6 (for $m = 3$ and 5, the corresponding processes take place with lower efficiencies, see Section 5.1).

$$Pt_4CH_2^+ + H_2O \quad \rightarrow \quad Pt_4CH_2O^+ + H_2 \tag{4.6}$$

Analogous reactions are expected to occur for Pt_4CHD^+ and $Pt_4CD_2^+$. However, experiment does not find an increase of $Pt_4CH_2O^+$ at the expense of $Pt_4CD_2^+$ (Figure 4.4). Exposure of mass-selected $Pt_4CH_2O^+$ to H_2 confirms the occurrence of the reverse process, reaction 4.7.

$$Pt_4CH_2O^+ + H_2 \quad \rightarrow \quad Pt_4CH_2^+ + H_2O \tag{4.7}$$

The resulting equilibrium is characterized by $K = k(4.6) / k(4.7) = 13 \pm 9$ and $\Delta_r G°(298 \text{ K}) = - 6 \pm 3$ kJ mol^{-1}. Thus, the thermodynamic equilibria connecting $Pt_4^+/CH_4/H_2O$, $Pt_4CH_2^+/H_2/H_2O$, and $Pt_4CH_2O^+/2\,H_2$ are fully established (Scheme 4.1).

$$Pt_4^+ + CH_4 + H_2O \underset{}{\overset{K = 1}{\rightleftharpoons}} Pt_4CH_2^+ + H_2 + H_2O \underset{}{\overset{K = 13}{\rightleftharpoons}} Pt_4CH_2O^+ + 2H_2$$

Scheme 4.1.

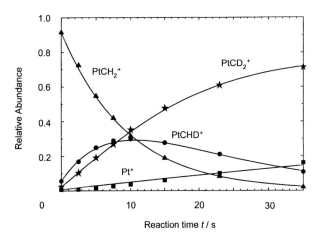

Figure 4.1. Observed (symbols) and modeled (solid lines) ion abundances in the reactions of PtCH$_2^+$ with D$_2$.

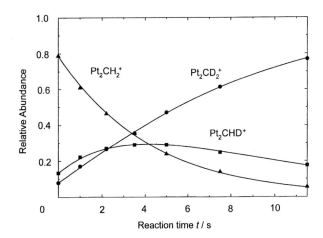

Figure 4.2. Observed (symbols) and modeled (solid lines) ion abundances in the reactions of Pt$_2$CH$_2^+$ with D$_2$.

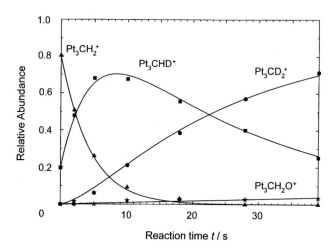

Figure 4.3. Observed (symbols) and modeled (solid lines) ion abundances in the reactions of $Pt_3CH_2^+$ with D_2.

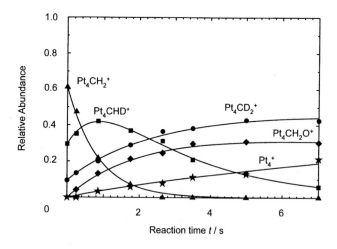

Figure 4.4. Observed (symbols) and modeled (solid lines) ion abundances in the reactions of $Pt_4CH_2^+$ with D_2.

In the case of the pentanuclear cluster $Pt_5CH_2^+$, reaction with background water is observed as well but not studied in full detail (see Section 5.1 for a comprehensive discussion). The intermediate product of stepwise exchange, Pt_5CHD^+, reaches a maximum relative abundance similar to those found for $PtCHD^+$ and Pt_2CHD^+.

The quantitative analysis is based on the kinetic scheme defined by the H/D exchange reactions 4.5a - c and reaction 4.4 for $m = 1$ and 4. Further, the reactions with background water are taken into account for $m = 3$ - 5. In order to reduce the flexibility of the model, identical rate constants are assumed for $Pt_mCH_2^+$, Pt_mCHD^+, and $Pt_mCD_2^+$ in the side reactions for a given m. This approximation corresponds to the postulation of negligible KIEs for these processes. In the case of reaction 4.4, rather small KIEs are found for the reverse dehydrogenation 4.3 (Table 4.1) such that the restriction imposed in the present model appears acceptable. Unfortunately, no independent data is available for the reactions with H_2O. Nonetheless, the model applied is considered reasonable as the imposition or suspension of the restrictions does not significantly affect the quality of the fit.

Table 4.2. Bimolecular rate constants k for H/D exchange reactions of $Pt_mCH_2^+$ with D_2, reactions 4.5a - c.

reaction	k / 10^{-10} cm^3 s^{-1} [a]				
	$m = 1$	2	3	4	5
4.5a	2.5 ± 0.3	2.1 ± 0.4	3.3 ± 0.4	4.1 ± 0.9	2.3 ± 0.8
4.5b	2.4 ± 0.5	2.1 ± 0.5	5.3 ± 0.5	9 ± 5	-[b]
4.5c	1.1 ± 0.3	1.1 ± 0.4	< 0.2	-[b]	2.3 ± 0.8

[a] Error bars given only account for relative uncertainties. [b] Data quality does not allow reliable quantification.

The results of the modeling essentially confirm the findings of the qualitative evaluation (Table 4.2). $PtCH_2^+$ and $Pt_2CH_2^+$ behave very similarly in all respects whereas $Pt_3CH_2^+$ shows a different reactivity, strongly favoring stepwise compared to direct exchange. For $m = 4$ and 5, the limited data quality in combination with the occurrence of side reactions lowers the accuracy of the derived rate constants. Thus, only for the first stepwise exchange, reaction 4.5a, a comparison of all different cluster sizes is feasible. Although all rate constants are rather similar, a slightly increased reactivity arises for $m = 4$. This finding reflects the lower stability established for $Pt_4CH_2^+$.

A further comparison also comprises the dehydrogenation of CH_2D_2 by Pt_m^+, reactions 4.3a - c. Reactions 4.3b and c yield the same products like reactions 4.5a and c, respectively. Whereas all four H/D atoms are *a priori* equivalent in the reactions involving CH_2D_2, the situation obviously is different for the reactions of $Pt_mCH_2^+$ with D_2. The degree of H/D equilibration achieved here may be inferred from comparison with the dehydrogenation reactions 4.5a and c. For complete equilibration, similar product distributions are expected for both reactions whereas the occurrence of significant discrepancies would indicate an incomplete H/D equilibration in the reaction of $PtCH_2^+$ with D_2. After compensation for statistical partitioning (1:4:1 for losses of H_2/HD/D_2 from $[Pt_m,C,H_2,D_2]^+$), one finds that the microscopic rate constant for reaction 4.5c (loss of H_2) is higher than that of reaction 4.5a (loss of HD) by a factor of 1.8 ± 0.6 and 2.1 ± 0.9 for $m = 1$ and 2, respectively. These values are somewhat higher than the KIEs derived from reactions 4.3b and c (Table 4.1) but still agree within the error margins. Hence, H/D equilibration in the course of reactions 4.5a and c appears likely for $m = 1$ and 2. The apparently slightly higher ratios for H_2 loss (reaction 4.5c) might be rationalized by differences in the internal energies in the collision complexes. Compared to the reaction of Pt_m^+ with CH_2D_2, reactions 4.5a and c correspond to the energetically lower entrance channel. For the resulting decreased effective temperatures, differences in the zero-point energies become more important such that elimination of H_2 is expected to be favored more strongly, as found in experiment. In contrast, the ratio between the statistically corrected rate constants for reactions 4.5c and a is < 0.3 in the case of the trimer. This value clearly is inconsistent with the KIE determined for reaction 4.3b and c and, thus, suggests an incomplete H/D equilibration for this cluster size. Seemingly, the details of the systems' PESs distinctly depend on the very cluster size.

4.4 Energy-Dependent CID of $Pt_mCH_2^+$

Whereas the above experiments address the whole Pt_m^+/CH_4 systems, energy-dependent CID of $Pt_mCH_2^+$ promises to provide additional insight with respect to the Pt_m^+/CH_2

subsystems. For mononuclear PtCH$_2^+$, loss of CH$_2$ is the predominant process, reaction 4.8.

$$Pt_mCH_2^+ \quad \rightarrow \quad Pt_m^+ + CH_2 \tag{4.8}$$

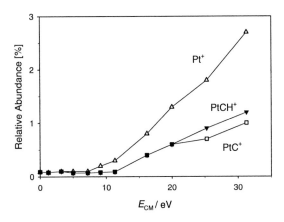

Figure 4.5. Relative abundances of the fragment ions in energy-dependent CID of PtCH$_2^+$.

From the present experiments applying argon as collision gas, an appearance energy $AE = 5 - 10$ eV (500 - 1000 kJ mol^{-1}) can be derived (Figure 4.5). This value is roughly consistent with the apparent threshold between 5 and 6 eV reported by Zhang *et al.* using guided-ion beam mass spectrometry with xenon as collision gas and thus demonstrates the qualitative validity of the approach pursued in the present work.[37] Losses of H and H$_2$, reactions 4.9 and 4.10, are observed to a significantly lesser extent for $m = 1$. Because these reactions are thermochemically less demanding than cleavage of the Pt$^+$–CH$_2$ bond ($\Delta_r H_0^\circ(4.9, m = 1) = 346 \pm 10$ and $\Delta_r H_0^\circ(4.10, m = 1) = 260 \pm 6$ *versus* $D_0(Pt^+–CH_2) = 463 \pm 3$ kJ mol^{-1}),[37] their low efficiencies point to the operation of considerable barriers.

$$Pt_mCH_2^+ \quad \rightarrow \quad Pt_mCH^+ + H \tag{4.9}$$

$$\rightarrow \quad Pt_mC^+ + H_2 \tag{4.10}$$

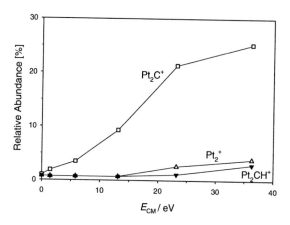

Figure 4.6. Relative abundances of the fragment ions in energy-dependent CID of $Pt_2CH_2^+$.

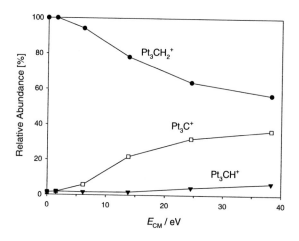

Figure 4.7. Relative abundances of $Pt_3CH_2^+$ and its fragment ions in energy-dependent CID.

In marked contrast, dehydrogenation according to reaction 4.10 predominates for $m = 2$ (Figure 4.6). The low appearance energy of this process, $AE \leq 1$ eV, accounts for its relatively high efficiency whereas reactions 4.8 and 4.9 require elevated collision energies and are much less effective. For the larger clusters, carbide formation 4.10 gains even more in importance, with appearance energies $AE < 2$ eV. In the case of $m = 3$, the only other fragmentation observed is reaction 4.9 giving small amounts of Pt_3CH^+ (Figure 4.7). For $Pt_4CH_2^+$ and $Pt_5CH_2^+$, this dissociation channel no longer occurs whereas, in addition to simple dehydrogenation, one and, for $m = 4$, two platinum atoms are lost at elevated collision energies, reactions 4.11 and 4.12, respectively (Figures 4.8 and 4.9).

$$Pt_mCH_2^+ \quad \rightarrow \quad Pt_{m-1}C^+ + [Pt,H_2] \qquad (4.11)$$

$$Pt_4CH_2^+ \quad \rightarrow \quad Pt_2C^+ + [Pt_2,H_2] \qquad (4.12)$$

The finding that $Pt_4CH_2^+$ does only give platinum-carbide clusters Pt_4C^+, Pt_3C^+, and Pt_2C^+, but no bare Pt_4^+ upon CID demonstrates a strong binding of the carbon atom to the cluster core. Most probably, this interaction involves more than a single metal center for all cluster sizes, thereby leading to larger bond-dissociation energies than that of the mononuclear carbide, $D_0(Pt^+–C) = 524 \pm 5$ kJ mol^{-1}.[37] A similar situation was found for Fe$_m$C$^+$ clusters.[157]

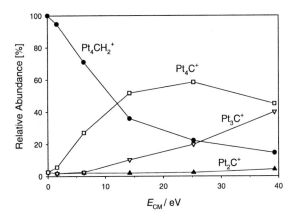

Figure 4.8. Relative abundances of $Pt_4CH_2^+$ and its fragment ions in energy-dependent CID.

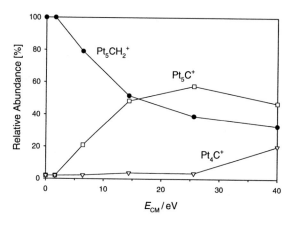

Figure 4.9. Relative abundances of $Pt_5CH_2^+$ and its fragment ions in energy-dependent CID.

4.5 Inferences from Experiment

So far, structural aspects have not been considered at all. *A priori* four different structural types appear possible for the $Pt_mCH_2^+$ cluster ions, Scheme 4.2.

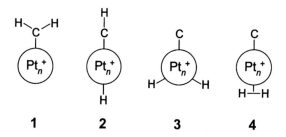

Scheme 4.2.

Type **1** can be referred to as a carbene structure whereas **2** represents a carbyne-hydrido species. Type **3** and **4** both correspond to carbide structures, **3** with two hydrido ligands and **4** with one adsorbed dihydrogen molecule. For mononuclear $PtCH_2^+$, theory predicts the carbene species **1** as global minimum of the system's PES.[35,36] Do analogous structures apply to the clusters as well? The facile elimination of H$_2$ from collisionally activated Pt$_m$CH$_2^+$, $m = 2 - 5$, at first sight seems to suggest structure **4** for these ions. However, other experimental findings are inconsistent with this assignment. First, loss of atomic H, reaction 4.9, observed for $m = 1 - 3$, appears highly improbable for ions with an intact, relatively weakly coordinating H$_2$ molecule. Further, the ready occurrence of stepwise H/D exchange upon exposure of Pt$_m$CH$_2^+$ to D$_2$ provides evidence against structure **4** as well because direct substitution of H$_2$ by D$_2$, reaction 4.5c, is supposed to largely prevail for this type. The stepwise exchange reactions 4.5a and b also render the existence of carbyne-hydrido species **2** unlikely. This species contains two non-equivalent hydrogen atoms that should exhibit significantly different tendencies towards substitution; however, quite similar rate constants are derived for reactions 4.5a and b (Table 4.2).

Discrimination between the remaining structures **1** and **3** is more difficult. Whereas the facile loss of H$_2$ observed upon CID supports structural type **3**, one would expect **3** to also display an enhanced reactivity with respect to H/D exchange. Yet, the rate constants obtained for the clusters are comparable to those for mononuclear PtCH$_2^+$ whose carbene structure was proven theoretically.[37] In addition, the KIEs associated with the reactions of Pt$_m^+$ with CH$_2$D$_2$ do not show marked differences between Pt$^+$ and the larger clusters, thus not pointing to a change in structure either. On the basis of these findings, presence of platinum-carbene clusters **1** appears slightly more probable than of carbide clusters **3**. Presumably, a definite solution of this structural problem has to await major progress in the theoretical description of these systems or the availability of novel experimental methods. With regard to the latter, coupling of mass spectrometry with optical spectroscopy would be particularly helpful in the present context.[158-160] However, the considerable instrumental requirements of platinum-cluster generation itself and the rather low ion abundances achieved would render such enterprise a truly formidable task.

Concerning the reactivity of the Pt$_m$CH$_2^+$ ions, the question of the systems' global energetic minima loses relevance at any rate. The experiments demonstrate that H/D exchange processes readily take place and thus indicate low barriers associated with rearrangements. Therefore, different parts of the PES are accessible in the course of a chemical reaction. Low barriers with respect to C–H bond breaking can also be inferred from the small KIEs observed for the reactions of Pt$_m^+$ with CH$_2$D$_2$. Pt$^+$ and the homologous clusters essentially behave the same in this respect whereas they show some differences as far as thermodynamics are concerned. Compared to methane dehydrogenation by mononuclear Pt^+, the analogous processes for Pt_m^+, $m = 2$, 3, and 5, appear to be more strongly exothermic as the reverse reaction 4.4 does not occur for these cluster sizes. In contrast, just the opposite situation is found for the tetramer where less favorable thermodynamics decelerate dehydrogenation and enable the back reaction. It remains to be seen below, how far the decreased stability of $Pt_4CH_2^+$ affects its reactivity toward other substrates.

The most important difference between mononuclear platinum and the homologous clusters concerns H_2 loss from $Pt_mCH_2^+$. Whereas this process does not play a significant role for energetic $PtCH_2^+$, it predominates for the homologous cluster ions, presumably because of better stabilization of the resulting carbides by multi-center interactions. The possibility of H_2 elimination as a viable exit channel therefore also has to be taken into account in the further course of the present work probing the reactivities of $Pt_mCH_2^+$ clusters.

5   Reactivity of Platinum-Carbene and -Carbide Clusters

After exploration of the Pt_m^+/CH_4 systems and their potential-energy surfaces, the reactivity of the platinum-carbene clusters $Pt_mCH_2^+$ can be addressed. The underlying motif in these studies is the problem of methane functionalization and possible implications of the gas-phase model for heterogeneous catalysis. As a probe for the chemical behavior of the $Pt_mCH_2^+$ clusters, their reactions with dioxygen, methane, ammonia, and water are considered. NH_3 and H_2O are typical nucleophiles appropriate as examples in a study of carbon-heteroatom coupling. C–N bond formation represents the key step in the industrially important synthesis of hydrogen cyanide in both the DEGUSSA and the Andrussov process.[17-22] Platinum-mediated C–O coupling yielding methanol or formaldehyde has not been established in heterogeneous catalysis yet, though it would also be highly desirable. CH_4, on the other hand, is an element hydride as well but lacks nucleophilicity, such that a comparison between the reactivities of the different neutral reactants promises to reveal the influence of the substrates' electronic properties on the course of reaction. Moreover, C–C bond formation might compete with carbon-heteroatom coupling and lead to unwanted by-products such that elucidation of these processes is warranted too. The reaction of the clusters with O_2, the most versatile oxidant, provides another potential route for methane functionalization. As O_2 is also used together with CH_4 and NH_3 in the Andrussov process, knowledge on the reactivity of $Pt_mCH_2^+$ toward O_2 might be helpful in this respect as well.

According to the energy-dependent CID experiments discussed in Section 4.4, also the platinum-carbide clusters Pt_mC^+ correspond to accessible species within the Pt_m^+/CH_4 systems. Consequently, the reactivity of Pt_mC^+ ions is of interest in the present context as well and promises to complete the picture obtained from the reactions of $Pt_mCH_2^+$. Furthermore, the reactivities of the carbide clusters can also be contrasted with those of bare platinum clusters. Such a comparison should reveal the influence of the carbide ligand as a model for ad-atom effects. With regard to methane functionalization in heterogeneous catalysis, this approach obviously bears particular relevance to the problem of soot formation. Like in the case of $Pt_mCH_2^+$, the reactions of Pt_mC^+ with NH_3, H_2O, CH_4, and O_2 are considered.

5.1   Reactions of $Pt_mCH_2^+$ Clusters

The reactivity of the mononuclear carbene species $PtCH_2^+$ has previously been investigated extensively.[34-36,39,40,148] For comparison, these data are included in the present study.

Reactions with Ammonia.   Reaction of $PtCH_2^+$ with NH_3 occurs efficiently (Table 5.1), yielding three different products, reactions 5.1 - 5.3 with $m = 1$.[39,40]

$$PtCH_2^+ + NH_3 \quad \rightarrow \quad NH_4^+ + PtCH \quad\quad (5.1)$$

$$PtCH_2^+ + NH_3 \quad \rightarrow \quad CH_2NH_2^+ + PtH \quad\quad (5.2)$$

$$Pt_mCH_2^+ + NH_3 \quad \rightarrow \quad [Pt_m,C,N,H_3]^+ + H_2 \quad\quad (5.3)$$

Whereas reaction 5.1, a simple proton transfer from the platinum carbene to NH_3, accounts for only 5 % of the total product formation, reaction 5.2 constitutes the main product channel (70 % b.r.) and unambiguously proves C–N bond coupling. Interpretation of reaction 5.3 (25 % b.r.) is less straightforward because a structural assignment of the ionic product needs to be achieved at first.[39,40]

For the homologous clusters $Pt_mCH_2^+$, $m = 2 - 5$, only reaction 5.3 is observed (Table 5.1). The absence of reactions 5.1 and 5.2 for the cluster ions is no surprise because their occurrence in the case of $PtCH_2^+$ reflects the high *IE* of atomic Pt, $IE(\text{Pt}) = 9.0$ eV,[118] which disfavors location of the positive charge at the metal-containing product. For the clusters, however, the larger metal core can more easily accommodate a positive charge (compare $IE(\text{Pt}_2) = 8.68 \pm 0.02$ eV)[119] such that reaction 5.3 is expected to gain in importance, as observed experimentally. A similar situation was encountered in the case of the reactions of Pt_m^+ with CH_3NH_2, see Section 3.2.

Consideration of the consecutive reactions of $[Pt_m,C,N,H_3]^+$ provides further insight into the structures of these ions. The mononuclear $[Pt,C,N,H_3]^+$ species reacts with another NH_3 molecule under dehydrogenation, reaction 5.4.[39,40] In contrast, the cluster ions

$[Pt_m,C,N,H_3]^+$, $m = 2 - 5$, do not activate ammonia but simply form adducts, reaction 5.5 (with φ up to 8 % for $m = 5$). Presumably, these processes are assisted by termolecular stabilization.

$$[Pt,C,N,H_3]^+ + NH_3 \quad \rightarrow \quad [Pt,C,N_2,H_4]^+ + H_2 \qquad\qquad (5.4)$$

$$[Pt_m,C,N,H_3]^+ + NH_3 \quad \rightarrow \quad [Pt_m,C,N_2,H_6]^+ \qquad\qquad (5.5)$$

The different behavior observed for $[Pt,C,N,H_3]^+$ on one hand and for the clusters $[Pt_m,C,N,H_3]^+$ on the other might indicate the presence of distinct ion structures. DFT calculations predict $[Pt,C,N,H_3]^+$ to correspond to the aminocarbene complex $PtC(H)NH_2^+$.[40] In the consecutive reaction with NH_3 (reaction 5.4), the hydrogen atom bound to carbon is replaced with an amino group, thus generating the bisaminocarbene species $PtC(NH_2)_2^+$ as the product.[40] Experimentally, this hypothesis can be probed by means of isotopic labeling. As shown previously, $PtCD_2^+$ loses HD in the process analogous to reaction 5.3 for $m = 1$, reaction 5.3a. In the consecutive reaction 5.4a, again loss of HD occurs. Both observations are in full accordance with the structural assignments provided by theory.[39,40]

$$PtCD_2^+ + NH_3 \qquad\qquad \rightarrow \qquad [Pt,C,N,H_2D]^+ + HD \qquad\qquad (5.3a)$$

$$[Pt,C,N,H_2D]^+ + NH_3 \rightarrow \qquad [Pt,C,N_2,H_4]^+ + HD \qquad\qquad (5.4a)$$

In marked contrast, the deuterated carbenes of platinum clusters ions exclusively lose D_2 upon reaction with NH_3, reaction 5.3b with $m = 2 - 5$. This is clear evidence that the carbenes derived from platinum clusters behave differently than that containing only a single metal center. Specifically, aminocarbene structures appear unlikely for $[Pt_m,C,N,H_3]^+$, $m = 2 - 5$, because either selective loss of HD or H/D scrambling is expected, but not exclusive elimination of D_2. Instead, the labeling experiments suggest that the NH_3 molecule remains intact and forms an adduct with a platinum-carbide moiety, i.e., $Pt_mC(NH_3)^+$. The same conclusion can be inferred from CID of $[Pt_m,C,N,H_3]^+$, which yields Pt_mC^+ as the only ionic products at variable collisional energies, reaction 5.6. In contrast, CID of the aminocarbene complex $PtC(H)NH_2^+$ gives a mixture of $[Pt,C,N,H]^+$, $CH_2NH_2^+$, and Pt^+ as ionic fragments;[39] this finding, again, demonstrates the dissimilarity between the system $PtCH_2^+/NH_3$ and the formally analogous platinum-cluster ions.

$$Pt_mCD_2^+ + NH_3 \qquad\qquad \rightarrow \qquad [Pt_m,C,N,H_3]^+ + D_2 \quad (m = 2 - 5) \qquad (5.3b)$$

$$[Pt_m,C,N,H_3]^+ \quad \rightarrow \quad Pt_mC^+ + NH_3 \qquad (m = 2 - 5) \qquad (5.6)$$

Table 1. Bimolecular rate constants k and efficiencies φ for the reactions of $Pt_mCH_2^+$ with NH_3, H_2O, CH_4, and O_2, respectively.

reaction	m	$k / cm^3 s^{-1}$	φ
$Pt_mCH_2^+ + NH_3 \rightarrow NH_4^+ + Pt_mCH$	1	$3.1 \times 10^{-11\,a}$	0.015
$Pt_mCH_2^+ + NH_3 \rightarrow CH_2NH_2^+ + Pt_mH$	1	$4.3 \times 10^{-10\,a}$	0.21
$Pt_mCH_2^+ + NH_3 \rightarrow [Pt_m,C,H_3,N]^+ + H_2$	1	$1.6 \times 10^{-10\,a}$	0.089
	2	9.7×10^{-10}	0.49
	3	9.6×10^{-10}	0.48
	4	1.7×10^{-9}	0.86
	5	1.2×10^{-9}	0.61
$Pt_mCH_2^+ + H_2O \rightarrow Pt_mCO^+ + 2\,H_2$	1	$2 \times 10^{-13\,b}$	8×10^{-5}
$Pt_mCH_2^+ + H_2O \rightarrow [Pt_m,C,H_2,O]^+ + H_2$	1	$4 \times 10^{-12\,b}$	2×10^{-3}
	2	$\leq 8 \times 10^{-14}$	$\leq 3 \times 10^{-5}$
	3	1.1×10^{-11}	5×10^{-3}
	4	9.4×10^{-10}	0.41
	5	1.0×10^{-10}	0.044
$Pt_mCH_2^+ + CH_4 \rightarrow [Pt_m,C_2,H_4]^+ + H_2$	1	$2 \times 10^{-11\,a}$	0.02
	2 - 4	$\leq 2 \times 10^{-12}$	$\leq 2 \times 10^{-3}$
	5	2.5×10^{-10}	0.26
$Pt_mCH_2^+ + O_2 \rightarrow Pt_mO^+ + CH_2O$	1	$7 \times 10^{-12\,c}$	0.012
$Pt_mCH_2^+ + O_2 \rightarrow Pt_m^+ + [C,H_2,O_2]$	1	$1.6 \times 10^{-11\,c}$	0.028
	2	$\leq 2 \times 10^{-13}$	$\leq 4 \times 10^{-4}$
	3	4.5×10^{-12}	8×10^{-3}
	4	9.0×10^{-11}	0.17
	5	1.3×10^{-10}	0.25

[a] Taken from ref. [40]. [b] Taken from ref. [148]. [c] Taken from ref. [34].

Reactions with Water. Upon changing the substrate from ammonia to water, a drastic decrease in reactivity is observed for $PtCH_2^+$ (Table 5.1). Apparently, the lower nucleophilicity of H_2O renders an attack at the electrophilic carbene less favorable than in the case of NH_3. The different electronic properties of H_2O are also mirrored in a change of product channels, reactions 5.7 and 5.8 with $m = 1$.[148]

$$PtCH_2^+ + H_2O \quad \rightarrow \quad PtCO^+ + 2\,H_2 \qquad (5.7)$$

$$Pt_mCH_2^+ + H_2O \quad \rightarrow \quad [Pt_m,C,H_2,O]^+ + H_2 \qquad (5.8)$$

Whereas reaction 5.7 only represents a minor process (5 % b.r.), reaction 5.8 is the main channel yielding the hydroxycarbene $PtC(H)OH^+$ for $m = 1$ in analogy to the coresponding reaction with NH_3.[148] The two other processes observed for the system $PtCH_2^+/NH_3$, reactions 5.1 and 5.2, do not have counterparts in the case of water. Most likely, the lower basicity of H_2O and the increased electronegativity of oxygen result in an unfavorable thermochemistry with respect to charge transfer away from the metal center. Similarly to $PtC(H)NH_2^+$, $PtC(H)OH^+$ undergoes an efficient consecutive reaction with a further substrate molecule, reaction 5.9.[148]

$$PtC(H)OH^+ + H_2O \quad \rightarrow \quad [Pt,C,H_2,O_2]^+ + H_2 \qquad (5.9)$$

Regarding the cluster ions, $Pt_2CH_2^+$ does not react with H_2O, whereas the larger clusters $Pt_mCH_2^+$, $m = 3 - 5$, afford dehydrogenation, reaction 5.8. As seen above, this finding alone is not sufficient to discriminate between the two reactivity patterns identified so far. However, the fact that no measurable consecutive reactions occur at the pressures applied is a first indication that the clusters behave differently from $PtCH_2^+$.[161] Again, deuterium labeling, reaction 5.8a, combined with CID experiments, reaction 5.10, provides more specific information.

$$Pt_mCD_2^+ + H_2O \quad \rightarrow \quad [Pt_m,C,H_2,O]^+ + D_2 \qquad (5.8a)$$

$$[Pt_m,C,H_2,O]^+ \quad \rightarrow \quad Pt_mC^+ + H_2O \qquad (5.10)$$

Upon reaction of $Pt_mCD_2^+$, $m = 3 - 5$, with H_2O, D_2 is lost exclusively (reaction 5.8a). This result is in marked contrast to elimination of HD in the reaction of $PtCD_2^+$ with H_2O,[148] whereas it parallels the clusters' reactivities toward NH_3. Moreover, CID of the $[Pt_m,C,H_2,O]^+$ products solely gives Pt_mC^+ as ionic fragment, reaction 5.10. Much like in the case of ammonia, these results thus indicate the formation of mere adducts between platinum-carbide clusters and H_2O, *i.e.*, $Pt_mC(H_2O)^+$. Note that this finding is fully in line with CID of $Pt_mCH_2^+$ yielding carbide clusters, see section 4.4. Apparently, H_2 elimination is the preferred exit channel for energetic $Pt_mCH_2^+$, $m = 2 - 5$, irrespective of the way the energy is supplied.

Further information can be obtained from a comparison of the rate constants of the different reactions observed (Table 5.1). The large efficiencies in the reactions with NH_3 for all cluster sizes studied imply that the encounter of both reactants releases sufficient energy to easily overcome all barriers involved en route to the products. In contrast, the lower basicity of water reduces the interaction energy such that differences in the reactivities of different cluster sizes become apparent. Whereas no reaction occurs for the dinuclear carbene $Pt_2CH_2^+$, the tri- and pentanuclear species react with moderate and the tetranuclear carbene with high efficiency. This order is supposed to inversely correlate with the stabilities of the carbene clusters. Indeed, the most reactive species, $Pt_4CH_2^+$, is the one with the lowest stability, see Section 4.2.

Reactions with Methane. As the third element hydride included in the present study, methane differs from its counterparts NH_3 and H_2O in its lack of a lone electron pair and the resulting absence of nucleophilicity. Only $PtCH_2^+$ and $Pt_5CH_2^+$ induce dehydrogenation of a further CH_4 molecule, reaction 5.11 with $m = 1$ and 5 (Table 5.1).[39]

$$Pt_mCH_2^+ + CH_4 \quad \rightarrow \quad [Pt_m,C_2,H_4]^+ + H_2 \qquad (5.11)$$

Compared to $Pt_mCH_2^+$, $m = 2$ - 4, the distinct reactivity of the mononuclear carbene can be explained by the larger number of open coordination sites in a single Pt atom which are obviously still not all occupied after addition of the first CH_2 fragment; in the product, the coordination number is somewhat reduced again because both methylene entities combine to one ethylene molecule.[40] Presumably, this decrease in the coordination number also accounts for slow further sequential addition/dehydrogenation reactions up to $PtC_5H_{10}^+$.[33] On the other hand, the pentamer appears large enough that addition of the first CH_2 fragment does not alter the geometric and electronic structures of the cluster drastically so that reaction with a further methane molecule can readily occur at another Pt center ($\varphi = 0.26$, Table 3.2). Achatz *et al.* also reported adsorption of a third CH_4 resulting in $Pt_5C_3H_8^+$, this reaction presumably being accompanied by termolecular stabilization.[57] As expected, such simple association reactions become more important for larger clusters and higher pressures.[43,58]

To further probe the structure of the $[Pt_5,C_2,H_4]^+$ product, this ion is subjected to CID which affords single and double dehydrogenation, but no losses of C_1 or C_2 fragments. If both methylene fragments combined to yield an ethylene molecule coordinated to the Pt_5^+ core, facile ligand exchange (ex) would be expected upon exposure of isotopically labeled $[Pt_5,C_2,D_4]^+$ to C_2H_4 ($k_{ex} \approx \frac{1}{2} k_c$, k_c = collision rate). Because such exchange is not observed above the noise level ($k_{ex} \leq \frac{1}{4} k_c$), a $Pt_5(C_2X_4)^+$ (X = H, D) structure appears less likely. This conclusion supports the hypothesis of separated Pt centers in Pt_5^+ reacting with CH_4. With regard to its reactivity, $[Pt_5,C_2,D_4]^+$ adds H_2O concomitant with loss of D_2, reaction 5.12; an analogous process occurs in the presence of NH_3, reaction 5.13. Accordingly, formal addition of a further CH_2 fragment to $Pt_5CH_2^+$ does not seem to significantly change its chemical behavior. Again, this finding is not compatible with a re-combination of both methylene moieties in the pentanuclear cluster.

$$[Pt_5,C_2,D_4]^+ + H_2O \quad \rightarrow \quad [Pt_5,C_2,D_2,H_2,O]^+ + D_2 \qquad (5.12)$$

$$[Pt_5,C_2,D_4]^+ + NH_3 \quad \rightarrow \quad [Pt_5,C_2,D_2,H_3,N]^+ + D_2 \qquad (5.13)$$

Reactions with Dioxygen. Another potential way to methane functionalization relies on reaction with oxygen. $PtCH_2^+$ was found to slowly react with O_2 (Table 5.1), yielding PtO^+ (30 % b.r.) and Pt^+ (70 % b.r.) as ionic products according to reactions 5.14 and 5.15 with $m = 1$.[34]

$$PtCH_2^+ + O_2 \quad \rightarrow \quad PtO^+ + CH_2O \qquad (5.14)$$

$$Pt_mCH_2^+ + O_2 \quad \rightarrow \quad Pt_m^+ + [C,H_2,O_2] \qquad (5.15)$$

Whereas the formation of formaldehyde in reaction 5.14 can safely be inferred on thermochemical grounds, the nature of the neutral product formed in reaction 5.15 is less clear. For the mononuclear system, elaborate theoretical studies indicate that formation of either HCOOH, CO/H_2O, or CO_2/H_2 is feasible and that most likely a mixture of these species is generated.[36] Taking into account the consecutive reactions occurring for Pt^+ and PtO^+ in the presence of CH_4 and O_2, an extended catalytic cycle for methane oxidation evolves. Note, however, that the only moderate selectivities of the different processes somewhat impair its attractiveness.[34,36]

With respect to the carbene clusters $Pt_mCH_2^+$, no reactivity is observed for $m = 2$, whereas reaction 5.15 takes place with increasing efficiencies for the larger cluster ions. In the case of $Pt_5CH_2^+$, small amounts of Pt_5O^+ are observed as well which might point to the occurrence of a process analogous to reaction 5.14 as a minor channel for $m = 5$. This assignment is uncertain, however, because the inevitable presence of traces of background water in the high-vacuum system also leads to formation of $Pt_5C(H_2O)^+$ (reaction 5.8) which, in analogy to $Pt_4C(H_2O)^+$, in turn might afford Pt_5O^+ under exposure to O_2 (see below). While similarly to the $PtCH_2^+/O_2$ system the structure of the neutral product of reaction 5.15 cannot be inferred from experiment, formation of HCOOH, CO/H_2O, or CO_2/H_2 appears reasonable. Obviously, the reactions of the clusters $Pt_mCH_2^+$, $m = 3 - 5$, with O_2 mediate carbon-oxygen bond formation, in contrast to the reactions with H_2O. Whereas the $Pt_mCH_2^+$ ions only lose their hydrogen atoms in the latter processes, it comes as no surprise that the reactions with O_2 also involve the carbon atom. Combination with the two hydrogen atoms merely saturates two valences of O_2 and, thus, would lead to thermochemically unfavorable products such as H_2O_2 or H_2O/O. In contrast, participation of carbon allows reduction of both oxygen atoms to their most stable oxidation state –II, thereby providing a strong driving-force for the overall reaction.

Simultaneous Reactions with Nucleophiles and Dioxygen. The distinct reactivities of the platinum-carbene clusters toward O_2 also provoke an investigation into the corresponding reactions of $Pt_mC(NH_3)^+$ and $Pt_mC(H_2O)^+$. However, as the production of these ions requires leaking-in of NH_3 and H_2O, respectively, into the ICR cell, the presence of these substrates besides O_2 significantly obscures the kinetic analysis of the reactions of interest. Therefore, the derivation of quantitative data for these processes is not indicated, and only the cluster size $m = 4$ is studied as an example. Upon reaction of $Pt_4C(NH_3)^+$ with O_2, $[Pt_4,H,N]^+$ is formed as primary product, reaction 5.16, which then gives rise to a plethora of consecutive reactions (*inter alia* yielding $[Pt_4,H_2,N]^+$ and $[Pt_4,H,N,O]^+$, whose origin is uncertain).

$$Pt_4C(NH_3)^+ + O_2 \quad \rightarrow \quad [Pt_4,H,N]^+ + [C,H_2,O_2] \tag{5.16}$$

The neutral product of reaction 5.16 has the same formula like the one of reaction 5.15 such that the same structural possibilities may apply. With respect to the ionic product, assumption of an imine species, *i.e.*, Pt_4NH^+, appears reasonable. In any case, the reaction is clear evidence for oxidation of NH_3. As this process certainly requires surpassing of considerable barriers, its occurrence implies the availability of significant amounts of energy during the course of the reaction. Whereas loss of NH_3 concomitant with formation of CO_2 appears as a viable alternative process from a thermochemical point of view, this particular reaction does not take place in experiment. The very reactivity observed indicates a strong binding between NH_3 and the cluster core. Lacking other information, the binding energy of the mononuclear system, $D_0(Pt^+–NH_3) = 274 \pm 12$ kJ mol^{-1},[113] may be used for comparison. The presence of more than a single metal center in the corresponding clusters is thought to further increase the interaction energy between NH_3 and the cluster core and may also help to reduce the barriers associated with NH_3 activation.

Reaction of $Pt_4C(H_2O)^+$ with O_2 yields $[Pt_4,C,O_2]^+$ as the primary product, reaction 5.17. Pt_4O^+ and Pt_4^+ are observed as consecutive products, but their origin cannot be inferred unambiguously.

$$Pt_4C(H_2{}^{16}O)^+ + {}^{18}O_2 \;\rightarrow\; [Pt_4,C,{}^{16}O,{}^{18}O]^+ + H_2{}^{18}O \qquad\qquad (5.17)$$

Reaction 5.17 formally corresponds to a substitution of H_2O for O_2. However, the isotopic labeling applied proves an activation of the water ligand in the course of the reaction, which requires extensive bond rearrangements. In contrast to reaction 5.16, only H_2O instead of $[C,H_2,O_2]$ is lost. If one assumes comparable pathways for both reactions, $[Pt_4,C,O_2]^+$ should correspond to a platinum-oxide cluster ligated by one CO molecule.[162] Apparently, the process involving $Pt_4C(H_2O)^+$ does not release sufficient energy to expel the CO ligand whereas such an elimination is possible for the $Pt_4C(NH_3)^+/O_2$ system.

5.2   Reactions of Pt_mC^+ Clusters

Unlike the situation for the platinum carbenes, neither the behavior of the smallest carbide, PtC^+, nor those of the homologous clusters has been investigated before. Pt_mC^+ ions can be generated by reaction of Pt_m^+ with CH_4 (see Section 2.6) or, for $m = 1$, with C_3O_2 at elevated energies.[99,100]

Reactions with Ammonia.   The reactions observed for the system PtC^+/NH_3 are rather complex. As primary processes, both dehydrogenation (65 % b.r.) and loss of a single hydrogen atom (35 % b.r.) take place, reaction 5.18 with $m = 1$ and reaction 5.19 (Table 5.2).

$$Pt_mC^+ + NH_3 \quad \rightarrow \quad [Pt_m,C,H,N]^+ + H_2 \tag{5.18}$$

$$PtC^+ + NH_3 \quad \rightarrow \quad [Pt,C,H_2,N]^+ + H \tag{5.19}$$

Reaction 5.19 is remarkable because it involves the release of atomic hydrogen, which is a highly energetic species. The occurrence of this process implies that PtC^+ must lie energetically high as well, resulting in enhanced reactivity and reduced selectivity. Concerning reaction 5.18, the most interesting question is whether or not this process mediates C–N bond coupling. CID, as potential means to probe an ion's connectivity, is not considered a proper method in the present case because loss of HCN (or HNC) is expected to be the energetically most favorable fragmentation channel of $[Pt,C,H,N]^+$ regardless of its structure. Some more information can be gained from the consecutive reactions. The major secondary product NH_4^+ arises from both $[Pt,C,H,N]^+$ and $[Pt,C,H_2,N]^+$ by simple proton transfer, whereas the minor consecutive product $PtNH_3^+$ is attributed to ligand exchange of $[Pt,C,H,N]^+$, reaction 5.20. Modeling of the product distributions on the basis of this kinetic scheme reproduces the experimental data quite well.

$$[Pt,C,H,N]^+ + NH_3 \quad \rightarrow \quad PtNH_3^+ + [C,H,N] \tag{5.20}$$

Occurrence of reaction 5.20 suggests the presence of a pre-formed HCN or HNC unit in $[Pt,C,H,N]^+$. Thus, PtC^+ appears to induce C–N bond formation upon reaction with NH_3 and resembles the carbene $PtCH_2^+$ in this respect. Most probably, PtC^+/NH_3 and $PtCH_2^+/NH_3$ are different entrance channels to the same PES. The latter corresponds to an early step on the way towards HCN/HNC formation and stops after single dehydrogenation with the aminocarbene $PtC(H)NH_2^+$ as product (besides generation of $CH_2NH_2^+$). Further elimination of H_2 requires additional energy, $i.e.$, CID of $PtC(H)NH_2^+$. In contrast, PtC^+/NH_3 enters the potential-energy surface at a later point, as loss of the first H_2 equivalent has already occurred in comparison with $PtCH_2^+/NH_3$. The energy gained from the interaction energy between NH_3 and PtC^+ is sufficient for dehydrogenation.

For the carbide derived from the platinum dimer, Pt_2C^+, reaction 5.18 is the only primary process observed. Compared to $m = 1$, the lower efficiency of this reaction (Table 5.2), and also the absence of neutral open-shell products suggest Pt_2C^+ to be a significantly less energetic species than its smaller homologue. Nevertheless, its ability to activate NH_3 demonstrates a distinct reactivity of this species. Consideration of the consecutive reactions helps to elucidate the connectivity of the $[Pt_2,C,N,H]^+$ product. In contrast to its smaller analogue, $[Pt_2,C,N,H]^+$ does not undergo ligand exchange with NH_3 (reaction 5.20) but simply adds this substrate, probably assisted by termolecular stabilization. This process does not necessarily indicate the pre-formation of a HCN or HNC unit in $[Pt_2,C,N,H]^+$ because ligand exchange might be expected in that case. Moreover, protolysis reactions giving NH_4^+ do only play a minor role for this system whereas they predominate for the mononuclear homologue. As HCN or HNC bound to a metal cation are supposed to exhibit an appreciable acidity, the low tendency of $[Pt_2,C,N,H]^+$ towards proton transfer does in fact dispute the presence of an HCN or HCN entity as well. However, if C–N coupling does not take place, the question arises which effect the carbide ligand has in NH_3 activation at all. Interestingly, bare Pt_2^+ even more readily dehydrogenates NH_3, see Section 3.2. Although the strong binding of the carbon atom certainly significantly alters the electronic and geometric features of the cluster's metal core, the dehydrogenation observed might still reflect the intrinsic reactivity of the platinum dimer.

$$Pt_mC^+ + NH_3 \quad \rightarrow \quad [Pt_m,C,N,H_3]^+ \tag{5.21}$$

For the larger clusters, simple addition of NH_3 takes place as primary process, reaction 5.21 with $m = 3 - 5$. The efficiency of this reaction is low for $m = 3$ but increases with cluster size (Table 5.2). As discussed above, such a behavior is characteristic of association reactions in the highly diluted gas phase where collisional and radiative stabilization of the encounter complexes gain in importance for larger systems. Simple adduct formation leads to $Pt_mC(NH_3)^+$ structures, *i.e.*, those inferred for the products of the reactions between $Pt_mCH_2^+$ and NH_3, reaction 5.3. Similar to $PtCH_2^+$ and PtC^+, the reactions of NH_3 with both the platinum-carbene and carbide clusters access the same PESs. Accordingly, $Pt_mC(NH_3)^+$ adds further NH_3, reaction 5.5, regardless of its origin from $Pt_mCH_2^+$ or Pt_mC^+.

The reactivities of Pt_mC^+, $m = 3 - 5$, toward ammonia may also be compared with those of the bare platinum clusters Pt_m^+, $m = 3 - 5$ (see above for $m = 2$), which simply add NH_3 as well. Given the rather low specificity of this reaction type, these parallels in behavior alone do not prove similar reactivities for Pt_m^+ and Pt_mC^+, however. Remarkable features of the bare clusters Pt_4^+ and particularly Pt_5^+ are the N–H bond activations observed in consecutive reactions with NH_3, see Section 3.1. Notably, the same behavior is found for

$Pt_5C(NH_3)^+$. Reaction with NH_3 yields $[Pt_5,C,N_2,H_6]^+$ and $[Pt_5,C,N_2,H_4]^+$, reactions 5.22 (70 % b.r.) and 5.23 (30 % b.r.). Apparently, the reactivities of the platinum-carbide clusters towards NH_3 indeed resemble those of $Pt_m{}^+$. The carbide ligand seems to act as an ad-atom which only slightly perturbs the reactivity of the cluster core.

$$Pt_mC(NH_3)^+ + NH_3 \quad \rightarrow \quad [Pt_5,C,N_2,H_6]^+ \tag{5.22}$$

$$Pt_mC(NH_3)^+ + NH_3 \quad \rightarrow \quad [Pt_5,C,N_2,H_4]^+ + H_2 \tag{5.23}$$

Reactions with Water. In contrast to the diverse reactivity of Pt_mC^+ clusters towards NH_3, no reactions at all are observed with water. Again, this difference is attributed to the lower nucleophilicity of H_2O which also manifests itself in the extremely low reaction rates for the formation of adducts with bare $Pt_m{}^+$ clusters, see Section 3.1.

Table 5.2. Bimolecular rate constants k and efficiencies φ for the reactions of Pt_mC^+ with NH_3, CH_4, and O_2, respectively.

reaction	m	$k \,/\, \mathrm{cm^3\,s^{-1}}$	φ
$Pt_mC^+ + NH_3 \rightarrow [Pt_m,C,H,N]^+ + H_2$	1	3.7×10^{-10}	0.18
	2	7.2×10^{-11}	0.036
$Pt_mC^+ + NH_3 \rightarrow [Pt_m,C,H_2,N]^+ + H$	1	1.9×10^{-10}	0.095
$Pt_mC^+ + NH_3 \rightarrow [Pt_m,C,H_3,N]^+$	3	$1.5 \times 10^{-11\,a}$	8×10^{-3}
	4	$8.7 \times 10^{-11\,a}$	0.044
	5	$1.7 \times 10^{-10\,a}$	0.086
$Pt_mC^+ + CH_4 \rightarrow [Pt_m,C_2,H_2]^+ + H_2$	1 - 4	$\leq 2 \times 10^{-12}$	$\leq 2 \times 10^{-3}$
	5	5.7×10^{-10}	0.60
$Pt_mC^+ + O_2 \rightarrow Pt_m{}^+ + CO_2$	1	3.9×10^{-12}	7×10^{-3}
$Pt_mC^+ + O_2 \rightarrow Pt_mO^+ + CO$	1	6.7×10^{-12}	0.012
	2	$\leq 1 \times 10^{-13}$	$\leq 2 \times 10^{-4}$
	3	$\leq 5 \times 10^{-13}$	$\leq 1 \times 10^{-3}$
	4	1.6×10^{-10}	0.30
	5	2.8×10^{-10}	0.53
$Pt_mC^+ + O_2 \rightarrow Pt_mCO^+ + O$	1	3.4×10^{-12}	6×10^{-3}
$Pt_mC^+ + O_2 \rightarrow Pt_{m-1}C^+ + PtO_2$	4	4.2×10^{-11}	0.079

[a] Apparent bimolecular rate constant at $p(NH_3) \approx 3 \times 10^{-8}$mbar.

Reactions with Methane.   The similarities in the reactivities of Pt_m^+ and Pt_mC^+ clusters entirely disappear in the case of CH_4, however. Whereas carbene formation occurs for all bare Pt_m^+ clusters (see Section 3.2), only Pt_5C^+ is able to dehydrogenate methane, reaction 5.24.

$$Pt_5C^+ + CH_4 \quad \rightarrow \quad [Pt_5,C_2,H_2]^+ + H_2 \tag{5.24}$$

This situation clearly resembles that observed for $Pt_mCH_2^+$, where also all clusters except the pentanuclear ion fail to activate CH_4. In contrast to its carbide counterpart, however, the carbene derived from atomic platinum mediates slow dehydrogenation of CH_4. Presumably, this difference arises from the distinct bond orders in $PtCH_2^+$ and PtC^{+}.[37] Whereas the carbene exhibits a double bound, the binding in the carbide rather corresponds to a triple bond. Thus, in PtC^+ the valences of the metal center are supposed to be more saturated which would account for the decreased reactivity toward CH_4.

Reactions with Dioxygen.   Finally, the reactions of platinum-carbide clusters with O_2 are investigated. For mononuclear PtC^+, three different processes occur, reactions 5.25 - 5.27 with $m = 1$.

$$PtC^+ + O_2 \quad \rightarrow \quad Pt^+ + CO_2 \tag{5.25}$$

$$Pt_mC^+ + O_2 \quad \rightarrow \quad Pt_mO^+ + CO \tag{5.26}$$

$$PtC^+ + O_2 \quad \rightarrow \quad PtCO^+ + O \tag{5.27}$$

Reactions 5.25 (28 % b.r.) and 5.26 (48 % b.r.) are oxidation processes that meet expectations. At first sight surprising is the occurrence of reaction 5.27 (24 % b.r.) in which atomic oxygen is released. To render the overall process exoergic, the reactants are supposed to provide considerable amounts of energy. Thus, again, a relatively low stability of PtC^+ is implied. Of course, formation of the extremely stable C–O bond shifts the energy balance in favor of the products as well. The binding between the metal and the CO ligand as a whole ($D_0(Pt^+–CO) = 212 \pm 10$ kJ mol^{-1})[163] also significantly contributes to the achieved total exoergicity $\Delta_r H°(5.27) = -270 \pm 11$ kJ mol^{-1}.[164,165]

In contrast, Pt_2C^+ and Pt_3C^+ do not react with O_2. For Pt_4C^+ and Pt_5C^+, however, oxidation according to reaction 5.26 occurs quite efficiently, whereas losses of neutral CO_2 or O do not take place. Absence of the latter channel is in line with an increased stability of the platinum-carbide clusters compared to PtC^+. With respect to the trade-off between production of CO and CO_2, the former already prevails for atomic platinum; enlargement of the metal core is supposed to increase its oxophilicity and, thus, to further favor CO formation, in agreement with experiment.

Compared to the reactions between $Pt_mCH_2^+$ and O_2, oxidation of the carbides does not stop at the bare platinum clusters but proceeds until Pt_mO^+, which simply reflects the availability of two reduction equivalents less than in the case of the carbenes. Besides formation of CO, O_2 also induces degradation of the Pt_4C^+ cluster, reaction 5.28 (20 % b.r.). This type of reactivity corresponds to that known for Pt_m^+ (see Section 3.3) and, again, indicates similarities between the carbide and the bare platinum clusters.

$$Pt_4C^+ + O_2 \quad \rightarrow \quad Pt_3C^+ + PtO_2 \tag{5.28}$$

5.3   Implications with Respect to Heterogeneous Catalysis

Having studied the reactions of gaseous $Pt_mCH_2^+$ and Pt_mC^+ clusters in detail, these processes can be compared with the corresponding reactions of bulk platinum catalysts. In the case of oxygen, all platinum carbenes $Pt_mCH_2^+$ except the dinuclear system mediate C–O bond formation which reflects platinum's high catalytic activity in oxidation processes. As a drawback of this enhanced reactivity, the reaction does not stop at the stage of methanol but proceeds further to formation of CO and H_2O (or related compounds, see Section 5.1). This situation resembles that of heterogeneous catalysis where over-oxidation of alkanes is a notorious problem. Regarding controlled oxidation of methane, no commercial process based on O_2 as oxidant has been realized so far. Therefore, it is no surprise that similar difficulties are encountered in the gas-phase model.

The reactivitiy of $Pt_mCH_2^+$ toward the element hydrides CH_4, NH_3, and H_2O shows a clear correlation with the substrates' nucleophilicities. Whereas NH_3 as the strongest nucleophile reacts with high efficiencies, the reactivities of $Pt_mCH_2^+$ toward H_2O are significantly diminished; only in the case of $Pt_4CH_2^+$, the decreased stability of this specific cluster allows a fast reaction. Concerning CH_4, the absence of a free electron pair in this substrate and the resulting lack of nucleophilicity prevent any reaction with $Pt_mCH_2^+$, $m = 2$ - 4. At first sight, the relatively efficient reaction observed for $Pt_5CH_2^+$ seems to contradict this line of reasoning. However, a more thorough examination suggests participation of a second platinum center in this reaction without involving the first CH_2 entity. Obviously, the different mechanism of this process does not allow a reasonable comparison with the reactions employing NH_3 or H_2O. Hence, C–C coupling does not appear to be efficient for platinum clusters such that it cannot compete with the reaction involving NH_3. This selective reactivity should be highly welcome if C–N bond formation is intended, as is in the industrially important DEGUSSA process.

However, the experiments also demonstrate that the reactions of platinum-carbene clusters $Pt_mCH_2^+$, $m = 2$ - 5, with NH_3 and H_2O do not accomplish carbon-heteroatom bond formation but instead lead to carbide species. Thus, the reactivity of the cluster ions toward nucleophilic substrates strikingly differs from that of mononuclear $PtCH_2^+$ which was shown to mediate carbon-heteroatom coupling indeed.[39,40] Paradoxically, the simpler model seems to mimic the real catalytic process better than the more refined one based on cluster ions. Yet, one must not forget that the desired C–N coupling is not the one and only reaction occurring under the conditions of the DEGUSSA process. Similarly to other reaction systems in heterogeneous catalysis, this process suffers from formation of soot as undesired by-product.[166,167] Soot formation does not only waste methane as reactant but also causes catalyst deactivation. Generation of the carbide clusters from Pt_mC^+/NH_3 can be considered as a gas-phase equivalent of these reactions. Hence, only together, Pt^+ and Pt_m^+, $m \geq 2$, cover the most important processes observed for CH_4/NH_3 on heterogeneous platinum catalysts. The strong effect of cluster size on reactivity found for the gas-phase models might indicate the possibility of a similar influence of the microscopic surface texture in the DEGUSSA process as well. In the industrial process, platinum is applied as thin coating on tubular alumina supports. During activation of the catalyst, major reconstruction of the surface occurs and leads to the formation of a Pt-Al alloy.[168] It may therefore well be the case that the catalytically active sites contain only a few platinum atoms. In light of the present gas-phase experiments, even single Pt centers might be supposed to exhibit particularly high catalytic activities.

Regarding catalyst deactivation by soot formation, further insight can be obtained from the experiments that directly probe the reactivities of Pt_mC^+ clusters. Towards NH_3, platinum-carbide clusters behave almost identically to the corresponding bare Pt_m^+ ions, $m = 2$ - 5. Obviously, the carbon atom does not actively participate in the reactions taking place. Moreover, the close similarities between Pt_mC^+ and Pt_m^+ suggest comparable geometric and electronic properties for both types of clusters. In the case of O_2, oxidation processes directly involve the carbon atom of the carbide clusters and, thus, naturally do not have counterparts on the side of the bare platinum clusters. However, cluster degradation as the predominant process for Pt_m^+, to some extent also applies to Pt_mC^+, thereby indicating parallels between both cluster types in this respect as well. In contrast, Pt_m^+ and Pt_mC^+ strongly differ in their reactivities toward CH_4. Whereas all Pt_m^+ clusters, $m \leq 5$, dehydrogenate CH_4, Pt_5C^+ is the only platinum carbide found to undergo the analogous reaction. Transferred to the heterogeneous functionalization of alkanes, this observation would suggest considerable catalyst deactivation already at rather low surface coverage by soot. The strong influence of carbon ad-atoms on the clusters' reactivities toward CH_4 could result from the rather weak thermodynamic driving-force for the formation of $Pt_mCH_2^+$ carbene species. With CH_4 dehydrogenation by Pt_m^+ being approximately thermoneutral for $m = 1$ and 4 and presumably not very exothermic for the other cluster sizes either, subtle changes in the system, such as the presence of a carbide ligand, might well cease reactivity.

Like in the case of the platinum-carbene clusters, mononuclear PtC^+ behaves differently from its larger homologues. The reactivity of PtC^+ is particularly distinguished by the occurrence of products lying relatively high in energy, thus proposing a decreased stability for PtC^+ itself as well. *Vice versa*, the absence of the analogous products in the reactions of the larger clusters with $m \geq 2$ points to enhanced stabilities of Pt_mC^+, $m = 2 - 5$; the same conclusion can be drawn more directly from the energy-dependent CID of $Pt_mCH_2^+$, see Section 4.4. As a consequence of the strong Pt_m^+–C interaction for $m \geq 2$, C–N coupling is no longer feasible for the clusters. Interestingly, such a decrease in reactivity associated with the transition from mono- to polynuclear systems appears to be a general phenomenon for transition-metal compounds M_mX^+. Whereas FeO^+, for instance, is highly reactive and even succeeds in oxidation of CH_4,[169] its larger homologue $Fe_2O_2^+$ shows a significantly dampened reactivity.[170,171] Similarly, only $FeNH^+$ mediates insertion of the NH fragment into a C–H bond of benzene,[172] the dinuclear species Fe_2NH^+ being essentially unreactive.[173] Most probably, the interaction of the fragment X with more than a single metal center in the cluster ions enforces binding and thereby disfavors a transfer of X to other substrates. An analogous decrease in reactivity can be assumed for fragments X bound to extended metal surfaces.

In the further course of the present work, special attention is paid to the question whether C–N bond formation can be accomplished by appropriate tuning of the platinum clusters' reactivity. On the basis of the present results, one would expect that weakening of the metal-carbon interaction is necessary to achieve this goal. However, the example of the carbide clusters has already highlighted the risk of deactivation with respect to CH_4 dehydrogenation as the essential first step of the overall process; presumably, only a delicate balance in reactivity can provide the desired effect. For modulation of the system's reactivity, two different approaches are pursued that also have direct counterparts in heterogeneous catalysis. First, the development of simple gas-phase models for supported platinum catalysts is attempted. Thereafter, the influence of a second transition metal besides platinum is addressed.

6   Gas-Phase Models of Supported Platinum Catalysts

In heterogeneous catalysis, the catalytically active transition metal rather often is not used in pure form but loaded on a support. This approach allows to increase the catalyst's dispersion and, thus, to enlarge its active surface. Moreover, the support material affects the electronic properties of the catalytically active metal, thereby altering the catalyst's reactivity. Hence, variation of the support is an important means to control the catalytic performance. In the case of platinum catalysts, for instance, a correlation between the catalytic activity with respect to alkane combustion and the acid strength of the support could be established.[174,175] As was demonstrated, the effect of the support operates via changing the oxidation state of platinum.[175] A direct influence of the support on the energy of platinum's valence orbitals was also inferred from X-ray spectroscopy data.[176]

Gas-phase studies could provide a helpful complement to these studies. The idealized environment of the gas phase should make a comparison between the reactivities of platinum on different supports particularly revealing. Furthermore, these experiments would also help to bridge the gap between gaseous bare platinum ions and supported heterogeneous catalysts. Despite of their great potential, gas-phase models of supported catalysts have so far neither been developed for platinum nor for any other transition metal. The reason for this poor state of knowledge lies in the substantial difficulties associated with the synthetic realization of such models. Alumina and silica, as the most important support materials, are hard solids with high melting points and negligible vapor pressures such that their transfer into the gas phase by simple evaporation obviously is impossible. Likewise, laser ionization/vaporization of catalyst/support mixtures does not appear viable either, because this method proceeds via initial atomization and, thus, destruction of composite materials. In the following, two alternative strategies for the generation of $PtAlO_2^+$ and $PtSiO_2^+$ as potential gas-phase models of platinum supported on alumina or silica, respectively, are probed.

6.1   Attempts at the Generation of PtAlO$_2^+$

Possible strategies applicable to the formation of gaseous PtAlO$_2^+$ comprise two steps. First, generation of bimetallic PtAl$_n^+$ clusters by laser vaporization/ionization of an appropriate alloy and subsequent supersonic expansion is intended. Afterwards, oxidation of PtAl$_n^+$ might *inter alia* yield PtAlO$_2^+$.

The binary system Pt/Al contains the phase Pt$_3$Al at the platinum-rich end of the phase diagram.[177] For the present purpose, a rather high content of platinum appears favorable in order to compensate for its lower tendency towards ionization, compared with aluminum (IE(Al) = 5.99 *versus* IE(Pt) = 9.0 eV).[118] Depending on the settings of the instrument, laser vaporization/ionization of Pt$_3$Al[178] yields either Pt$^+$ or Al$^+$ in good abundances. However, only extremely low amounts of cluster ions are produced for widely differing operating conditions of the cluster source. Possibly, the high porosity of the alloy target pressed from powdery Pt$_3$Al negatively affects its suitability to cluster generation in comparison to homogeneous metal samples.

Despite this obstacle, PtAl$^+$ can be produced in sufficient quantities to probe its reactivity toward the oxidants O$_2$ and N$_2$O. Exposure of PtAl$^+$ to oxygen leads to cluster degradation by loss of PtO$_2$, reaction 6.1 ($k = 8.8 \times 10^{-11}$ cm^3 s^{-1}, $\varphi = 0.16$).

$$\text{PtAl}^+ + \text{O}_2 \quad \rightarrow \quad \text{Al}^+ + \text{PtO}_2 \tag{6.1}$$

Similar to the analogous reaction of Pt$_2^+$, reaction 3.11a, loss of PtO$_2$ rather than separate species can be inferred on thermochemical grounds. Localization of the positive charge at the aluminum atom is consistent with its low IE. Note, however, that the putatively nobler platinum is oxidized. Again, this finding points to the higher stability of Pt(IV) compared to Pt(II) in PtO that alternatively might be formed concomitant with AlO$^+$. At any rate, the reaction with O$_2$ obviously does not provide the desired result.

In the primary reaction of PtAl$^+$ with N$_2$O, transfer of one oxygen atom occurs rather efficiently, reaction 6.2 ($k = 2.5 \times 10^{-10}$ cm^3 s^{-1}, $\varphi = 0.36$).

$$\text{PtAl}^+ + \text{N}_2\text{O} \quad \rightarrow \quad \text{PtAlO}^+ + \text{N}_2 \tag{6.2}$$

Only Al$^+$ and [Al,H$_2$,O$_2$]$^+$ are observed as consecutive products, the former resulting from a secondary reaction with N$_2$O, reaction 6.3.

$$\text{PtAlO}^+ + \text{N}_2\text{O} \quad \rightarrow \quad \text{Al}^+ + \text{PtO}_2 + \text{N}_2 \tag{6.3}$$

Apparently, a second oxygen transfer takes place but does not yield $PtAlO_2^+$ as stable product. Occurrence of the dissociation strongly suggests that transient $[Pt,Al,O_2]^+$ rather corresponds to a weakly bound $Al^+(PtO_2)$ complex than to a Pt(II) species ligated by an AlO_2^- moiety. As only the latter corresponds to a reasonable gas-phase model of alumina-supported platinum, even the hypothetical stabilization of transient $[Pt,Al,O_2]^+$ would not lead to the intended result. Hence, $PtAlO_2^+$ does not seem to be appropriate in the present context for fundamental reasons. Other models such as $PtAl_2O_3^+$ might better mimic the structure of the solid but are even more demanding in their synthesis.

The second consecutive product, *i.e.*, $[Al,H_2,O_2]^+$, presumably arises from a very efficient reaction with background water. Replacement of the platinum atom with H_2O leaves behind an $AlO(H_2O)^+$ or, more probably, $Al(OH)_2^+$ ion, reaction 6.4.

$$PtAlO^+ + H_2O \quad \rightarrow \quad [Al,H_2,O_2]^+ + Pt \qquad (6.4)$$

6.2   Gas-Phase Synthesis of $PtSiO_2^+$

Considering the problems encountered in the generation of $PtAl^+$ by laser ionization/vaporization, a different approach is pursued in the synthesis of $PtSiO_2^+$. Thanks to the high abundances easily obtained for Pt^+, this ion appears to be a good starting point for the gas-phase synthesis. In the next step, silicon is introduced via reaction with silane as a volatile silicon compound. In contrast to its smaller homologue CH_4, SiH_4 is efficiently dehydrogenated twice, reaction 6.5 ($k = 7.8 \times 10^{-10}$ cm^3 s^{-1}, $\varphi = 0.75$).

$$Pt^+ + SiH_4 \quad \rightarrow \quad PtSi^+ + 2\,H_2 \qquad (6.5)$$

The facile activation of SiH_4 gives rise to a manifold of consecutive reactions of which formation of $PtSi_2^+$ and $PtSi_2H_2^+$, reactions 6.6 and 6.7, respectively, are the most prominent ones. Whereas $PtSi_2^+$ seems to be a final product, $PtSi_2H_2^+$ undergoes further reactions not studied in detail.

$$PtSi^+ + SiH_4 \quad \rightarrow \quad PtSi_2^+ + 2\,H_2 \qquad (6.6)$$

$$\rightarrow \quad PtSi_2H_2^+ + H_2 \qquad (6.7)$$

Next, $PtSi_2^+$ is exposed to oxygen. The overall reaction proceeds rather slowly ($k = 1.8 \times 10^{-11}$ cm^3 s^{-1}, $\varphi = 0.032$) and yields Pt^+ (40 % b.r.) and $PtSiO^+$ (50 % b.r.) as main products, reactions 6.8 and 6.9, respectively.

$$PtSi_2^+ + O_2 \qquad \rightarrow \qquad Pt^+ + [Si_2,O_2] \qquad\qquad (6.8)$$

$$\rightarrow \qquad PtSiO^+ + SiO \qquad\qquad (6.9)$$

At longer reaction times, Pt^+ increases at the expense of $PtSiO^+$, presumably because the latter is subject to hydrolysis by background water. The probably substantial exothermicity associated with this process apparently causes immediate dissociation of the putative intermediate $[Pt,Si,H_2,O_2]^+$. Among the remaining reaction products observed in small quantities, *i.e.*, PtO^+, $PtSi^+$, $PtSi_2OH^+$, and $SiOH^+$, the latter two also indicate interferences of background water. The structure of $PtSiO^+$ is probed by CID that mainly affords Pt^+ (75 %) besides smaller amounts of PtO^+ and $PtSi^+$. This behavior agrees with, but does not prove the presence of a complex consisting of Pt^+ and SiO.

Finally, reaction of $PtSiO^+$ with N_2O efficiently yields the desired $PtSiO_2^+$ ion (70 % b.r.) along with $PtSi^+$ (30 % b.r.), reactions 6.10 and 6.11, respectively ($k = 3.8 \times 10^{-10}$ cm^3 s^{-1}, $\varphi = 0.54$ for the overall reaction).

$$PtSiO^+ + N_2O \qquad \rightarrow \qquad PtSiO_2^+ + N_2 \qquad\qquad (6.10)$$

$$\rightarrow \qquad PtSi^+ + O_2 + N_2 \qquad\qquad (6.11)$$

The observed consecutive formation of Pt^+ is ascribed to hydrolysis of $PtSiO_2^+$ by background water, reaction 6.12. Deliberate addition of H_2O accelerates the generation of Pt^+ according to an approximate rate constant $k = 7 \times 10^{-10}$ cm^3 s^{-1} ($\varphi = 0.3$) for hydrolysis.

$$PtSiO_2^+ + H_2O \qquad \rightarrow \qquad Pt^+ + [Si,H_2,O_3]^+ \qquad\qquad (6.12)$$

Most probably, the neutral product of this reaction corresponds to metasilicic acid $SiO(OH)_2$ which is about 370 kJ mol^{-1} more stable than SiO_2/H_2O.[179,180] This immense difference in thermochemical stabilities reflects silicon's well-known preference of forming single rather than double bonds to oxygen.[130] In turn, the presence of an SiO_2 entity in $PtSiO_2^+$ might not be considered energetically favorable either. Notwithstanding, CID of $PtSiO_2^+$ exclusively leads to the loss of SiO_2 and, thus, does suggest a $Pt(SiO_2)^+$ structure for the parent ion. This connectivity may be viewed as analogous to silica-supported platinum (see Section 6.3, however).

In praxi, the gas-phase synthesis starts with pulsing-in SiH_4 to mass-selected and thermalized $^{195}Pt^+$. After a reaction delay of 2.5 s, a mixture of O_2 and N_2O is pulsed into the reaction cell for ca. 5 s and pumped off within further 5 s.[181] The synthesis is terminated by mass selection of $PtSiO_2^+$. For achieving sufficient abundances of the product ion, a careful adjustment of the pressures of the reactant gases proves essential.

6.3   Reactivities of $PtSiO_2^+$ and Pt^+ Toward Hydrocarbons

The reactivity of $PtSiO_2^+$ is probed by studying its reactions with methane, propane, 1-butene, and 1,4-cyclohexadiene. For comparison, also the reactions of bare Pt^+ with these substrates are investigated. Because the latter reactions proceed very efficiently, they are likely to interfere with the time-demanding synthesis of $PtSiO_2^+$ if the substrates are leaked in permanently. The application of low substrate pressures is not feasible either, because this approach would not allow monitoring of slow reactions between $PtSiO_2^+$ and the hydrocarbons. To circumvent this problem, the substrates are pulsed in only after completion of the synthesis of $PtSiO_2^+$ and pumped off prior to the next cycle; this procedure probes the reactivity of $PtSiO_2^+$ in a qualitative manner. However, in the case of 1,4-cyclohexadiene, also a quantitative analysis appears interesting, such that this substrate has to be leaked in continuously. All products arising from the reaction of Pt^+ with C_6H_8 during the synthesis of $PtSiO_2^+$ are ejected prior to the kinetic measurement.

In contrast to bare Pt^+, $PtSiO_2^+$ does not react with CH_4. This finding is not too surprising because one well expects SiO_2 to partially saturate the open coordination sites of Pt^+ and thereby decrease its reactivity. Yet, the magnitude of this effect remains to be established. How far has the C–H bond in the substrate to be weakened for activation by $PtSiO_2^+$ to occur? In the case of propane, pulsing-in the substrate only yields the adduct $[Pt,Si,O_2,C_3,H_8]^+$. This finding demonstrates the efficiency of termolecular stabilization at elevated pressures during the gas pulse, but does not indicate C–H bond activation. Instead, such an activation is evident for bare Pt^+ that induces rapid dehydrogenation of C_3H_8, reactions 6.13 and 6.14 ($k = 5.7 \times 10^{-10}$ cm^3 s^{-1}, $\varphi = 0.69$ for the overall reaction). The predominance of double dehydrogenation, reaction 6.13, not observed for CH_4, probably reflects the possibility of internal stabilization of the C_3H_4 moiety by formation of multiple bonds.

$Pt^+ + C_3H_8$	$\rightarrow$	$PtC_3H_4^+ + 2\,H_2$	75 %	(6.13)
	$\rightarrow$	$PtC_3H_6^+ + H_2$	25 %	(6.14)

Consecutive reactions with propane involve further dehydrogenation producing mainly $PtC_6H_{10}^+$ with smaller amounts of $PtC_6H_8^+$ and $PtC_6H_{12}^+$.

Further facilitation of C–H activation can be achieved by introduction of a double bond. However, still the reaction of $PtSiO_2^+$ with 1-butene only leads to the association product $[Pt,Si,O_2,C_4,H_8]^+$. In contrast, Pt^+ does not only accomplish C–H, but also C–C bond cleavages, reactions 6.15 - 6.18. Again, a high efficiency of $\varphi = 0.66$ is observed for the overall reaction ($k = 6.8 \times 10^{-10}$ cm^3 s^{-1}). The primary products undergo consecutive reactions yielding $PtC_8H_x^+$, $x = 8, 10, 12, 14$, and 16, as well as $PtC_{12}H_x^+$, $x = 14$ and 16.

$Pt^+ + C_4H_8$	$\rightarrow$	$PtC_4H_4^+ + 2\,H_2$	30 %	(6.15)
	$\rightarrow$	$PtC_4H_6^+ + H_2$	25 %	(6.16)
	$\rightarrow$	$PtC_2H_2^+ + [C_2,H_6]$	35 %	(6.17)
	$\rightarrow$	$PtC_2H_4^+ + [C_2,H_4]$	10 %	(6.18)

Finally, the reactivities of $PtSiO_2^+$ and Pt^+ toward 1,4-cyclohexadiene are investigated. This substrate turns out to be sufficiently activated for rapid reaction with $PtSiO_2^+$ ($k = 1.5 \times 10^{-9}$ cm^3 s^{-1}, $\varphi = 1.5$; the physically reasonable maximum value $\varphi = 1$ lies slightly outside the assumed experimental uncertainty of 30 %). The major process observed corresponds to hydride abstraction yielding protonated benzene as ionic product, reaction 6.19. In addition, three further processes take place, reaction 6.20 - 6.22.

$PtSiO_2^+ + C_6H_8$	$\rightarrow$	$C_6H_7^+ + [Pt,Si,O_2,H]$	55 %	(6.19)
	$\rightarrow$	$C_6H_8^+ + [Pt,Si,O_2]$	15 %	(6.20)
	$\rightarrow$	$[C_4,H_7,Si,O]^+ + [Pt,C_2,O,H]$	15 %	(6.21)
	$\rightarrow$	$PtC_6H_6^+ + [Si,O_2,H_2]$	15 %	(6.22)

Clearly, the low *IE* of 1,4-cyclohexadiene, $IE(C_6H_8) = 8.82 \pm 0.02$,[118] accounts for the electron transfer in reaction 6.20. Spontaneous occurrence of this reaction indicates that addition of the SiO_2 moiety to Pt^+ does not significantly lower, if at all, platinum's *IE*, $IE(Pt) = 9.0$.[118] With respect to the two remaining processes, reaction 6.22 represents a formal dehydrogenation of the substrate that meets expectations. Obviously, the energy released from this reaction suffices to expel the SiO_2 moiety as well; thus, formation of separate neutral species SiO_2/H_2 seems likely. In contrast, reaction 6.21 is not readily interpreted. The sum formula assigned to the ionic fragment is the only one consistent with a high-resolution mass determination. For the corresponding neutral [Pt,C_2,O,H] species, one might speculatively assume CO/PtCH structures. The generation of CO may be favored because of its particular thermochemical stability, and also PtCH is a viable product in principle as its formation upon hydride transfer from $PtCH_2^+$ to NH_3 demonstrates, reaction 5.1. Notwithstanding the uncertainties with regard to the ionic and neutral products of reaction 6.21, the mere occurrence of this process implies an active participation of the SiO_2 entity. Apparently, the role of the SiO_2 moiety is far beyond that of a spectator ligand and thus to some extent questions the adequacy of $PtSiO_2^+$ as gas-phase model for supported platinum.

The reactions of bare Pt^+ with 1,4-cyclobutadiene ($k = 7.2 \times 10^{-10}$ cm^3 s^{-1}, $\varphi = 0.71$ for the overall reaction) largely resemble those observed for $PtSiO_2^+$. In analogy to $PtSiO_2^+$, Pt^+ induces hydride and electron transfer, reactions 6.23 and 6.24, respectively, as well as dehydrogenation, reaction 6.25. The main reaction channel leads to double dehydrogenation (reaction 6.26), however, which does not have a counterpart in the case of $PtSiO_2^+$. This difference once more demonstrates attenuation of platinum's reactivity by the SiO_2 entity.

$$Pt^+ + C_6H_8 \quad \rightarrow \quad C_6H_7^+ + PtH \qquad\qquad 10\ \% \qquad\qquad (6.23)$$

$$\rightarrow \quad C_6H_8^+ + Pt \qquad\qquad 5\ \% \qquad\qquad (6.24)$$

$$\rightarrow \quad PtC_6H_6^+ + H_2 \qquad\qquad 30\ \% \qquad\qquad (6.25)$$

$$\rightarrow \quad PtC_6H_4^+ + 2\,H_2 \qquad\qquad 55\ \% \qquad\qquad (6.26)$$

As a conclusion, the experiments demonstrate a dramatic decrease in reactivity associated with the change from Pt^+ to $PtSiO_2^+$. The magnitude of this effect points to a substantial interaction between Pt^+ and the SiO_2 moiety and the saturation of open coordination sites at the metal center. Assuming a similar situation for silica-supported platinum catalysts might be premature, however. The gas-phase model contains a single SiO_2 moiety, thus enforcing the formation of thermochemically unfavorable silicon-oxygen double bonds. A significant interaction with the Pt^+ center is likely to partially avoid this disadvantageous situation. In contrast, bulk silica is distinguished by a particular thermochemical stability which results from fourfold coordination of silicon and the exclusive presence of Si–O single bonds in the Si_xO_{2x} network. Consequently, the interactions between the support and the catalytically active metal are supposed to be much weaker than that between Pt^+ and SiO_2 in gaseous $PtSiO_2^+$.

Although the synthesis of $PtSiO_2^+$ represents an interesting example how gas-phase models of supported catalysts can be accessed, $PtSiO_2^+$ itself does not appear to mimic the behavior of silica-supported platinum in a satisfactory manner. More realistic gas-phase models should include a larger section of the Si_xO_{2x} framework. In this respect, Basset's approach of employing polysiloxy complexes as model compounds for bulk silica might form a promising starting point also for gas-phase chemistry.[182]

7   Reactivity of Pt$_m$Au$_n^+$ Clusters

An important strategy in catalysis does not apply a single transition metal only but relies on bimetallic systems.[55] Obviously, the combination of different transition metals greatly increases the number of potentially active catalysts and may result in entirely novel reactivity patterns. A well-known example of this approach is provided by cobalt-molybdenum alloys that have been shown to activate dinitrogen more efficiently than the presently used iron or ruthenium systems and that might therefore constitute the next generation of catalysts for ammonia synthesis.[183-188] This case is particularly remarkable because cobalt and especially molybdenum alone have rather poor activities with respect to ammonia formation.[187] Moreover, the discovery of the Co-Mo system can serve as prime example of a rational, knowledge-based catalyst development. As demonstrated by Jacobsen *et al.*, the catalytic activity of a specific metal surface with regard to N$_2$ conversion crucially depends on its nitrogen-atom adsorption energy $E_{ads}(N)$.[187] Whereas pure cobalt interacts with nitrogen so weakly that it hardly affords N$_2$ dissociation in the first reaction step, the binding of the N atom to molybdenum is too strong and thus disfavors the consecutive reaction with hydrogen. In contrast, Co-Mo alloys turn out to have just the right intermediate affinity for atomic nitrogen. Jacobsen *et al.* proposed the potential general relevance of such interpolation schemes for the rational design of bimetallic catalysts.[187]

The problem of methane functionalization forms an interesting test case for this concept. As shown in Chapter 5, the platinum-carbene clusters Pt$_m$CH$_2^+$ do not undergo coupling reactions with ammonia because of unfavorably strong metal-carbene interactions. In analogy to the bimetallic Co-Mo system for NH$_3$ synthesis and according to the proposal of Jacobsen *et al.*, the combination of platinum with an appropriate second metal could improve the clusters' activities in terms of C–N coupling. Specifically, addition of a later transition metal with a more closely filled *d*-shell might decrease the metal-carbene binding energy and thus facilitate the coupling step with ammonia. As platinum's right-hand neighbor in the periodic table, gold appears to be the natural choice for combination with platinum.

In the following, the reactivities of the bimetallic Pt$_m$Au$_n^+$ clusters, $m + n \leq 4$, toward CH$_4$, O$_2$, NH$_3$, and CH$_3$NH$_2$ are studied. Against the background of methane functionalization and C–N coupling, the investigation of CH$_4$ and NH$_3$ is straightforward. Similarly, the reactions of CH$_3$NH$_2$ promise to provide an alternative access to the potential-energy surfaces relevant to C–N bond formation and to thus yield interesting complementary information. The reactivities of the Pt$_m$Au$_n^+$ species toward O$_2$ are mainly included for the sake of comparison with the data obtained for the homonuclear platinum clusters. Moreover, the preparation of pure Pt$_m$Au$_n^+$ clusters involves reactions with O$_2$.

7.1 Generation and Purification of Pt$_m$Au$_n^+$ Clusters

Like in the case of the gas-phase models for supported platinum clusters (see Chapter 6), bimetallic clusters ions can be produced by gas-phase syntheses or by laser ionization/vaporization of appropriate alloys. The first approach was pursued by Jacobson and Freiser, for instance, who generated FeCo$_2^+$ upon reaction of Fe$^+$ with Co$_2$(CO)$_8$ followed by CID.[189] Barrow et al. successfully applied the second, more general method to the production of even trimetallic cluster ions.[190]

In analogy, laser ionization/vaporization of a 1:1 platinum-gold alloy yields a mixture of Pt$_m$Au$_n^+$ ions. The small mass difference between Pt and Au in conjunction with platinum's broad isotope pattern (gold is a pure element) results in isobaric interferences. For instance, ^{195}Pt^{197}Au$^+$ as the most abundant mixed dinuclear cluster is overlapped by ^{196}Pt$_2^+$ and ^{194}Pt^{198}Pt$^+$. The difference between the exact masses of these ions lies in the order of 0.001 amu such that they are no longer resolved at typical experimental conditions (Figure 7.1a). Obviously, ion-ejection techniques cannot be used for the selective removal of Pt$_2^+$ (see Chapter 2). However, ion separation is achieved by taking advantage of the ions' different reactivities towards O$_2$. Whereas Pt$_2^+$ is efficiently oxidized by pulsed-in O$_2$, see Section 7.3, PtAu$^+$ does not react and is left behind. After this chemical purification (Figure 7.1b),[191] the ^{195}Pt^{197}Au$^+$ isotopomer can be selected by conventional ion-ejection techniques (Figure 7.1c) and is available for further investigation.

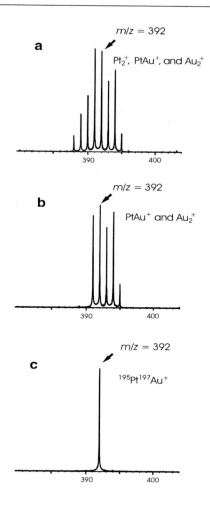

Figure 7.1. Mass range *m/z* = 380 - 404 (a) prior to reaction with O_2, (b) after reaction with O_2, and (c) after reaction with O_2 and ion ejection. The *m/z* values of the different isotopomers with significant abundances are 388 - 394 for Pt_2^+, 391, 392, 393, and 395 for $PtAu^+$, and 394 for Au_2^+.

Similar methods can be applied in the case of the gold-rich clusters $PtAu_2^+$ and $PtAu_3^+$. In contrast, Pt_3Au^+ and $Pt_2Au_2^+$ cannot be separated because of their apparently very similar reactivities toward the various substrates probed. As also the isotopic patterns of these clusters do not differ largely and thus do not allow their facile distinction either, the individual rate constants for their reactions cannot be determined. Notwithstanding, the rate constants for the different reactions investigated do not substantially vary for reactant ions with m/z = 782 - 784, although the predicted abundances of Pt$_3$Au$^+$ and Pt$_2$Au$_2^+$ significantly change in this mass range (100:83:54 for Pt$_3$Au$^+$ *versus* 40:79:100 for Pt$_2$Au$_2^+$).[192] To a first approximation, the reactions of Pt$_3$Au$^+$ and Pt$_2$Au$_2^+$ with the investigated substrates should therefore proceed with similar efficiencies. Thus, the individual rate constants are assumed to equal the combined values within a factor of 2.

7.2   Reactions of Pt$_m$Au$_n^+$ Clusters with CH$_4$

In contrast to Pt$_m^+$, pure gold clusters Au$_n^+$ do not dehydrogenate CH$_4$ nor undergo any other reaction with this substrate in the highly diluted gas phase. At higher pressures, association reactions can occur by termolecular stabilization (for $n \leq 15$). These processes are supposed to lead to inserted structures HAu$_n$CH$_3^+$.[193]

$$Pt_mAu_n^+ + CH_4 \qquad \rightarrow \qquad Pt_mAu_nCH_2^+ + H_2 \qquad\qquad (7.1)$$

Similar to Pt$_2^+$, the bimetallic PtAu$^+$ cluster activates CH$_4$, reaction 7.1 with $m = n = 1$ (Table 7.1). Pt$_2$Au$^+$ also induces dehydrogenation of CH$_4$, reaction 7.1 with $m = 2$ and $n = 1$, whereas PtAu$_2^+$ is unreactive. With respect to the tetranuclear bimetallic clusters, both Pt$_3$Au$^+$ and Pt$_2$Au$_2^+$ dehydrogenate CH$_4$ to afford the corresponding carbene complexes Pt$_m$Au$_n$CH$_2^+$ according to reaction 7.1. In contrast, the gold-rich tetramer ion PtAu$_3^+$ proves inert.

Reaction 7.1 is investigated in some more detail for PtAu$^+$. First, the labeled variant of this process is studied by application of CH$_2$D$_2$ as substrate. Taking into account statistical partitioning between the different products (1:4:1 for reactions 7.1a - c), intramolecular kinetic isotope effects KIE(7.1b/7.1a) = 0.9 ± 0.1 and KIE(7.1c/7.1b) = 1.3 ± 0.1 are derived. Thus, PtAu$^+$ much resembles Pt$_2^+$ and the larger Pt$_m^+$ clusters that show similarly small KIEs in the analogous processes (see Section 4.1). Like for the latter systems, the rate-limiting step of the overall reaction of PtAu$^+$ with CH$_4$ does obviously not involve C–H bond cleavage.

$$PtAu^+ + CH_2D_2 \quad \rightarrow \quad PtAuCH_2^+ + D_2 \tag{7.1a}$$

$$\rightarrow \quad PtAuCHD^+ + HD \tag{7.1b}$$

$$\rightarrow \quad PtAuCD_2^+ + H_2 \tag{7.1c}$$

Similarities between the $PtAu^+/CH_4$ and Pt_2^+/CH_4 systems are also inferred from CID experiments. In analogy to $Pt_2CH_2^+$, $PtAuCH_2^+$ predominantly loses H_2 under energetic collisions with argon. Additionally, losses of H and CH_2 occur to a much lesser extent.

Finally, the secondary reactions of the clusters ions with CH_4 are considered. $PtAuCH_2^+$ is found to be the only heteronuclear carbene that dehydrogenates a further molecule CH_4, reaction 7.2. Although the efficiency of this reaction is very low ($k = 6.7 \times 10^{-12}$ cm^3 s^{-1}, $\varphi = 7 \times 10^{-3}$), it nevertheless suggests that the carbene moiety in $PtAuCH_2^+$ might be available for coupling reactions. In this respect, $PtAu^+$ appears to resemble atomic Pt^+ more closely than Pt_2^+ or the larger homonuclear Pt_m^+ clusters, $m \leq 4$.

$$PtAuCH_2^+ + CH_4 \quad \rightarrow \quad [Pt,Au,C_2,H_4]^+ + H_2 \tag{7.2}$$

Table 7.1. Bimolecular rate constants k and efficiencies φ for the dehydrogenation reactions of $Pt_mAu_n^+$ clusters with CH_4, reaction 7.1.

m	n	k / cm^3 s^{-1}	φ
1	0	5.0×10^{-10}	0.51
0	1	$\leq 10^{-13}$	$\leq 10^{-4}$
2	0	8.2×10^{-10}	0.85
1	1	6.4×10^{-10}	0.67
0	2	$\leq 10^{-13}$	$\leq 10^{-4}$
3	0	6.0×10^{-10}	0.63
2	1	4.8×10^{-10}	0.50
1	2	$\leq 10^{-13}$	$\leq 10^{-4}$
0	3	$\leq 10^{-13}$	$\leq 10^{-4}$
4	0	1.5×10^{-11}	0.02
3	1	8.0×10^{-11} [a]	0.08
2	2		
1	3	$\leq 10^{-13}$	$\leq 10^{-4}$
0	4	$\leq 10^{-13}$	$\leq 10^{-4}$

[a] Combined value for Pt_3Au^+ and $Pt_2Au_2^+$, see Section 7.1 for details.

7.3 Reactions of $Pt_mAu_n^+$ Clusters with O_2

Unlike Pt_m^+ cluster ions (see Section 3.3), homonuclear gold clusters Au_n^+ are unreactive toward O_2.[193] Similarly, the smallest bimetallic cluster ion $PtAu^+$ does not react with O_2 either. Cluster degradation occurs if the cluster contains a second platinum atom like in Pt_2Au^+. In a process analogous to the reactions observed for homonuclear Pt_m^+ clusters, O_2 abstracts one platinum atom from the cluster core, reaction 7.3 with $m = 2$ and $n = 1$. In contrast, the Au-rich trinuclear cluster $PtAu_2^+$ does not react with O_2 (Table 7.2).

$$Pt_mAu_n^+ + O_2 \rightarrow Pt_{m-1}Au_n^+ + PtO_2 \qquad\qquad (7.3)$$

A parallel trend in reactivity evolves for the tetranuclear clusters. Whereas exposure to O_2 leads to fragmentation of both Pt_3Au^+ and $Pt_2Au_2^+$ according to reaction 7.3, the Au-rich cluster ion $PtAu_3^+$ proves inert. As expected, the primary product Pt_2Au^+ originating from the degradation of Pt_3Au^+ continues to react with O_2 under cluster degradation.

Table 7.2. Bimolecular rate constants k and efficiencies φ for the degradation reactions of $Pt_mAu_n^+$ clusters exposed to O_2, reaction 7.3.

m	n	k / cm^3 s^{-1}	φ
2	0	1.3×10^{-10}	0.23
1	1	$\leq 10^{-13}$	$\leq 10^{-4}$
0	2	$\leq 10^{-13}$	$\leq 10^{-4}$
3	0	7.6×10^{-13}	0.001
2	1	3.9×10^{-11}	0.07
1	2	$\leq 10^{-13}$	$\leq 10^{-4}$
0	3	$\leq 10^{-13}$	$\leq 10^{-4}$
4	0	1.4×10^{-10}	0.25
3	1	$2.4 \times 10^{-10\,a}$	0.43
2	2		
1	3	$\leq 10^{-13}$	$\leq 10^{-4}$
0	4	$\leq 10^{-13}$	$\leq 10^{-4}$

^{a} Combined value for Pt_3Au^+ and $Pt_2Au_2^+$, see 7.1 for details.

The reactions of bimetallic $Pt_mAu_n^+$ cluster ions with O_2 and CH_4 follow a clear rule. Clusters with a predominant fraction of platinum atoms undergo reactions analogous to those of homonuclear Pt_m^+ clusters, *i.e.*, they lose PtO_2 upon exposure to O_2 and they dehydrogenate CH_4. In contrast, clusters with a higher fraction of gold atoms are unreactive toward O_2 and CH_4, corresponding to the behavior of pure Au_m^+ clusters. For cluster ions containing equal amounts of platinum and gold, two cases can be distinguished. $PtAu^+$ reacts with CH_4 but not with O_2, thus exhibiting an intermediate reactivity. In contrast $Pt_2Au_2^+$ undergoes reactions with both CH_4 and O_2, such that platinum apparently determines the overall reactivity of this cluster.

A rationalization of these findings is facilitated by a comparison with the behavior of Pt^+ and Au^+ as the simplest constituents of the heteronuclear cluster ions. The fact that Au^+, unlike Pt^+, does not dehydrogenate CH_4 can be attributed to the lower thermochemical stability of $AuCH_2^+$ compared to $PtCH_2^+$, $D_0(Au^+–CH_2) \leq 372 \pm 3$ *versus* $D_0(Pt^+–CH_2) = 463 \pm 3$ kJ mol^{-1}.[37,194] This difference has been mainly ascribed to the high energy $E = 180$ kJ mol^{-1} required for the promotion of Au^+ from its $5d^{10}$ ground state to the excited $5d^9 6s^1$ state that is necessary for the formation of a double bond to the methylene fragment.[33] For the homologous Au_n^+ clusters, the electronic structures are much more complex than in the case of Au^+. Presumably, the presence of additional electrons leads to an increased valence saturation compared to the Pt_m^+ clusters which may account for the low reactivities of Au_n^+ clusters towards hydrocarbons, such as methane and benzene.[195] Note that the origin of the well-known catalytic activities of gaseous anionic or supported neutral gold clusters with respect to CO oxidation completely differs.[45-51] Here, it is not the substrate which transfers electron density into an empty orbital of the metal core, like in the reaction with CH_4;[193] in distinct contrast, the metal cluster acts as electron donor for adsorbed O_2.[49] Thus, an electropositive support or addition of an electron to the gaseous clusters is necessary to enable charge transfer to O_2, whereas the cationic clusters Au_n^+ do not react. In contrast, Pt_m^+ clusters are oxidized by O_2 despite their positive charge because of platinum's ability to adopt higher oxidation states than gold (+IV *versus* +III).

The reactivities of the heteronuclear $Pt_mAu_n^+$ clusters well fit into this scheme. Apparently, an increasing gold content saturates the clusters' valences but also diminishes their electron-donating properties, such that reactions with both CH_4 and O_2 no longer occur. For a strictly additive behavior, the bimetallic clusters should show a monotonic decrease of reactivities as a function of gold content. However, such a trend is only observed partly. Whereas the reaction rates of the dinuclear clusters meet expectations,

Pt$_2$Au$^+$ reacts significantly faster with O$_2$ than Pt$_3^+$ does. In the case of the tetranuclear clusters, the reactions of Pt$_3$Au$^+$ and Pt$_2$Au$_2^+$ with CH$_4$ are more efficient than for Pt$_4^+$, and also their reactions with O$_2$ are slightly faster than the oxidation of the pure Pt$_4^+$ cluster. Interestingly, exactly the reactions of Pt$_3^+$ with O$_2$ and Pt$_4^+$ with CH$_4$ are exceptionally inefficient compared to the analogous processes for other sizes of Pt$_m^+$ clusters ions, see Chapter 3. Hence, the anomalies appear to occur for the homonuclear Pt$_m^+$ ions rather than on the side of the mixed clusters. Presumably, the peculiar reactivities result from particular electronic or geometric features of Pt$_3^+$ and Pt$_4^+$, respectively, which are impaired by substitution of one platinum atom by gold. These size-specific effects obviously disturb the derivation of quantitative trends. Taking into account this situation, the composite reactivities of the Pt$_m$Au$_n^+$ clusters indeed appear to result from the individual reactivities of the components in a roughly additive manner.

7.3   Reactions of Pt$_m$Au$_n^+$ Clusters with NH$_3$

For the reactions of homonuclear Au$_n^+$ clusters with NH$_3$, two different types can be distinguished. Whereas slow association occurs for n = 1 and 3,[195] efficient cluster degradation takes place for n = 2 and 4, reaction 7.4 (Table 7.3). In the case of the tetramer, a second degradation process brings about loss of Au$_3$ (20 % b.r.), reaction 7.5. Given the much lower *IE* of the gold trimer, *IE*(Au$_3$) = 7.50 eV, compared to *IE*(Au) = 9.23 eV,[118,196] the charge distribution observed for the products of this reaction channel might appear surprising. Yet, AuNH$_3^+$ + Au$_3$ still is lower in energy than the alternative AuNH$_3$ + Au$_3^+$ asymptote because of the high bond-dissociation energy of the cationic gold-ammonia complex, D(Au$^+$–NH$_3$) = 297 ± 30 kJ mol^{-1} *versus* D(Au–NH$_3$) = 76 ± 6 kJ mol^{-1} (note that the most stable combination corresponding to Au$_3$NH$_3^+$ + Au is realized in reaction 7.4 with n = 4).[197,198] Essentially, relativistic effects account for the high stabilization of Au$^+$ by ammonia and other electron-donating ligands.[151,199]

$$Au_n^+ + NH_3 \quad \rightarrow \quad Au_{n-1}NH_3^+ + Au \qquad\qquad (7.4)$$

$$Au_4^+ + NH_3 \quad \rightarrow \quad AuNH_3^+ + Au_3 \qquad\qquad (7.5)$$

The fact that only Au$_2^+$ and Au$_4^+$ undergo cluster degradation whereas Au$_3^+$ is inert towards NH$_3$ in this respect, points to an even-odd alternation well-known for coinage-metal clusters in general[200] and gold-cluster ions in particular.[201-203] Having a $5d^{10}6s^1$ electron configuration, each Au atom added to the ionic cluster provides a single valence electron that is available for intra-cluster binding. Because electron pairing is energetically favorable, clusters with even numbers of electrons, *i.e.*, odd n for the cationic clusters, are more stable as demonstrated by a comparison of the bond-dissociation energies $D_0(Au_{n-1}^+-Au)$.[204,205]

Interestingly, the primary product of reaction 7.4 for $n = 4$, $Au_3NH_3^+$, is subject to a second degradation process, reaction 7.6, although the metal core has an even number of valence electrons and thus should bear an enhanced stability.

$$Au_3NH_3^+ + NH_3 \quad \rightarrow \quad Au(NH_3)_2^+ + Au_2 \qquad (7.6)$$

Possibly, the NH_3 ligand transfers an appreciable amount of electron density to the metal core and thereby causes a substantial perturbation of the simple even-odd alternation scheme such that it no longer adequately describes the cluster's electronic situation. Moreover, occurrence of reaction 7.6 might be attributed to the particular stability of the $Au(NH_3)_2^+$ cation which has previously been observed in the gas phase as well as in condensed matter (for the related case of $AuNH_3^+$, see above).[206,207] The anticipated loss of neutral Au_2 in reaction 7.6 (expulsion of two separate Au atoms would be hardly feasible thermochemically) equals the main dissociation channel for kinetically excited bare Au_n^+ cluster ions with odd values of n.[204]

With regard to the mixed $Pt_mAu_n^+$ clusters, $PtAu^+$ as the smallest representative reacts with NH_3 under cluster degradation in analogy to Au_2^+, reaction 7.7. Interestingly, the larger bimetallic cluster ions do not undergo corresponding degradation reactions but simply add NH_3 similarly to the homonuclear Pt_m^+ ions, reaction 7.8.

$$PtAu^+ + NH_3 \quad \rightarrow \quad PtNH_3^+ + Au \qquad (7.7)$$

$$Pt_mAu_n^+ + NH_3 \quad \rightarrow \quad Pt_mAu_nNH_3^+ \qquad (7.8)$$

Table 7.3. Apparent and bimolecular rate constants k and efficiencies φ for the reactions of $Pt_mAu_n^+$ clusters with NH_3.

products	m	n	k / cm^3 s^{-1}	φ
$Pt_mAu_nNH_3^+$	1	0	$5.0 \times 10^{-13\,a,b}$	3.0×10^{-4}
	0	1	$3.6 \times 10^{-14\,a}$	1.8×10^{-5}
	3	0	$5.4 \times 10^{-12\,a}$	3.0×10^{-3}
	2	1	$5.2 \times 10^{-12\,a}$	2.6×10^{-4}
	1	2	$\leq 10^{-12\,a}$	$\leq 10^{-3}$
	0	3	$\leq 10^{-12\,a}$	$\leq 10^{-3}$
	4	0	$2.3 \times 10^{-11\,a}$	0.012
	3	1	$3.0 \times 10^{-11\,a,c}$	0.015
	2	2		
	1	3	$1.0 \times 10^{-11\,a}$	5.0×10^{-3}
$Pt_mAu_nNH^+ + H_2$	2	0	5.4×10^{-10}	0.27
$Pt_mAu_{n-1}NH_3^+ + Au$	1	1	2.8×10^{-10}	0.14
	0	2	6.0×10^{-10}	0.30
	0	4	8.8×10^{-10}	0.46
$Pt_mAu_{n-3}NH_3^+ + Au_3$	0	4	2.2×10^{-10}	0.12

^{a} Apparent rate constants for $p(NH_3) \approx 10^{-7}$ - 10^{-6} mbar. ^{b} Taken from ref. [39]. ^{c} Combined value for Pt_3Au^+ and $Pt_2Au_2^+$, see Section 7.1 for details.

The efficiencies of the association reactions exhibit a marked dependence on cluster composition (Table 7.3). Whereas the reactions of Pt_2Au^+, Pt_3Au^+, and $Pt_2Au_2^+$ proceed with almost identical rates as those of the Pt_m^+ ions of equal cluster size, the Au-rich species show significantly decreased reactivities. This trend is consistent with the behavior of the bimetallic clusters towards CH_4 and O_2. Accordingly, the lower reactivities of the Au-rich clusters can be attributed to their increased electronic saturation which diminishes the interactions with the substrate. The close similarities between the Pt-rich $Pt_mAu_n^+$ and the homonuclear Pt_m^+ cluster ions are also evident from the consecutive reactions. As shown in Section 3.1, the larger ligated clusters $Pt_mN_yH_{3y}^+$ react with further NH_3 under dehydrogenation. The same behavior is observed in the case of $Pt_3Au^+/Pt_2Au_2^+$. For these cluster ions, two NH_3 molecules have to be added first before dehydrogenation occurs in the reaction with a third one.

7.4   Reactions of $Pt_mAu_n^+$ Clusters with CH_3NH_2

As outlined in Section 3.2, the reaction of atomic Pt^+ with CH_3NH_2 predominantly affords hydride abstraction (80 % b.r.) whereas single and double dehydrogenation processes are less efficient. Given that $D_0(Au–H) = 300$ kJ mol^{-1} $\approx$ $D_0(Pt–H) = 310$ kJ mol^{-1} and that $IE(Au) = 9.23$ eV $> IE(Pt) = 9.0$ eV,[118,165,208] hydride transfer should be energetically even more favored for the reaction of Au^+ with CH_3NH_2. Indeed, exposure of Au^+ to CH_3NH_2 exclusively results in the formation of protonated formimine and neutral AuH (as explicitly proven by using labeled CD_3NH_2), reaction 7.9 with $m = 0$ and $n = 1$.

$$Pt_mAu_n^+ + CH_3NH_2 \quad \rightarrow \quad CH_2NH_2^+ + Pt_mAu_nH \qquad (7.9)$$

The higher tendency of gold towards hydride abstraction is also evident in the case of the dinuclear clusters. Both the branching ratio and the efficiency of reaction 7.9 increase when going from Pt_2^+ via $PtAu^+$ to Au_2^+ (Table 7.4). Compared to atomic Pt^+ and Au^+, however, the fraction of hydride transfer is reduced for the dinuclear clusters in accordance with their presumably lower IEs. Whereas Pt_2^+ and $PtAu^+$ mainly accomplish double and, to a lesser extent, single dehydrogenation of CH_3NH_2, reactions 7.10 and 7.11, respectively, Au_2^+ does not undergo analogous processes, which resembles its inertness toward CH_4. Instead, substitution of one of the cluster's gold atoms by CH_3NH_2 leads to $AuCH_3NH_2^+$ which most probably corresponds to the complex of Au^+ with an intact methylamine ligand, reaction 7.12 with $n = 2$. This process parallels the NH_3-induced cluster degradation, reaction 7.3, fully consistent with the close similarity of NH_3 and CH_3NH_2.

$$Pt_mAu_n^+ + CH_3NH_2 \quad \rightarrow \quad [Pt_m,Au_n,C,H,N]^+ + 2\,H_2 \qquad (7.10)$$

$$\rightarrow \quad [Pt_m,Au_n,C,H_3,N]^+ + H_2 \qquad (7.11)$$

$$Au_n^+ + CH_3NH_2 \quad \rightarrow \quad Au_{n-1}CH_3NH_2^+ + Au \qquad (7.12)$$

The analogy between the reactivities of CH_3NH_2 and NH_3 also holds for the larger clusters studied. Whereas Au_3^+ does not undergo substitution according to reaction 7.12 (see below), this process represents the exclusive primary reaction channel in the case of $n = 4$. Apparently, the even-odd alternation already inferred from the reactions with NH_3 is quite a general feature of the reactivity of small cationic gold clusters. In addition, distinct similarities are also observed for the consecutive processes of the $Au_3NH_3^+/Au_3CH_3NH_2^+$

systems. In both cases, the reaction with a second substrate molecule leads to expulsion of neutral Au$_2$, reaction 7.6 *versus* 7.13.

$$Au_3CH_3NH_2^+ + CH_3NH_2 \quad \rightarrow \quad Au(CH_3NH_2)_2^+ + Au_2 \qquad (7.13)$$

Table 7.4. Bimolecular rate constants k and efficiencies φ for the reactions of Pt$_m$Au$_n^+$ clusters with methylamine.

products	m	n	k / 10^{-10} cm^3 s^{-1}	φ
CH$_2$NH$_2^+$ + Pt$_m$Au$_n$H	1	0	7.1	0.46
	0	1	7.3	0.47
	2	0	2.0	0.13
	1	1	3.3	0.22
	0	2	5.0	0.33
[Pt$_m$,Au$_n$,C,H$_5$,N]$^+$	0	3	1.4^a	0.10
[Pt$_m$,Au$_n$,C,H$_3$,N]$^+$ + H$_2$	1	0	0.52	0.03
	2	0	0.50	0.03
	1	1	0.73	0.05
	1	2	8.0	0.54
	0	3	2.1	0.14
[Pt$_m$,Au$_n$,C,H,N]$^+$ + 2 H$_2$	1	0	0.95	0.06
	2	0	7.5	0.50
	1	1	3.3	0.22
	3	0	14.0	0.92
	2	1	8.0	0.54
	4	0	10.0	0.68
	3	1	7.6^b	0.52
	2	2		
	1	3	7.6	0.52
[Pt$_m$,Au$_{n-1}$,C,H$_5$,N]$^+$ + Au	0	2	1.2	0.08
	0	4	8.2	0.56

^{a} Apparent rate constant for p(CH$_3$NH$_2$) $\approx 10^{-8}$ mbar. ^{b} Combined value for Pt$_3$Au$^+$ and Pt$_2$Au$_2^+$, see Section 7.1 for details.

The larger heteronuclear clusters exclusively undergo dehydrogenation processes with CH$_3$NH$_2$. Whereas PtAu$_2^+$ only affords single dehydrogenation, reaction 7.11 with $m = 1$

and $n = 2$, all other clusters eliminate two molecules H_2, reaction 7.10. The fact that even the Au-rich clusters achieve efficient dehydrogenation of CH_3NH_2 (Table 7.4) points to the activating effect of the amino group compared to non-substituted CH_4. According to this interpretation, one would assume the metal clusters to preferentially attack the methyl group of CH_3NH_2. In the case of the dinuclear $PtAu^+$ cluster, this issue is explicitly probed by means of isotopic labeling. Treatment of $PtAu^+$ with CD_3NH_2 yields $CD_2NH_2^+$, reaction 7.9a, along with $[Pt,Au,C,H,N]^+$, reaction 7.10a with $m = n = 1$ (single dehydrogenation in analogy to reaction 7.11 does not occur to an appreciable extent). Thus, the methyl group indeed forms the more reactive position in methylamine. Obviously, $PtAu^+$ behaves very similarly to Pt^+ and Pt_2^+ (see Section 3.2).

$$Pt_mAu_n^+ + CD_3NH_2 \quad\rightarrow\quad CD_2NH_2^+ + Pt_mAu_nH \qquad (7.9a)$$

$$\rightarrow\quad [Pt_m,Au_n,C,H,N] + D_2 + HD \qquad (7.10a)$$

In addition, also the high interaction energy between the cationic clusters and the dipolar amine is expected to further facilitate dehydrogenation. Although the enhanced reactivity of the substrate results in a leveling effect, some differences in the reaction patterns of the $Pt_mAu_n^+$ clusters still remain visible, namely for the trinuclear systems. As soon as Au predominates in the clusters' elemental composition, merely single dehydrogenation occurs. This finding is consistent with the above conclusions and the general picture of reactivity attenuation upon valence saturation. In addition to H_2 elimination, homonuclear Au_3^+ simply adds CH_3NH_2 (40 % b.r. for $p(CH_3NH_2) \approx 10^{-8}$ mbar) which further emphasizes the rather poor potential of pure cationic gold clusters with respect to bond activation.

Like in the case of the homonuclear Pt_m^+ clusters, the reactions of their bimetallic counterparts do not stop after dehydrogenation of the first substrate molecule but involve up to two further CH_3NH_2 units. For the last step of these reaction sequences, saturation effects manifest themselves in the change from double to single dehydrogenation. A different kind of consecutive process is only observed in the $PtAu^+/CH_3NH_2$ system. Here, both primary dehydrogenation products react with a further CH_3NH_2 molecule under elimination of AuH, reactions 7.14 and 7.15. Given the isolobal relation between AuH and H_2, these processes are best understood as aura-analogues of ordinary dehydrogenation reactions such as 7.11.[199]

$$[Pt,Au,C,H_3,N]^+ + CH_3NH_2 \quad \rightarrow \quad [Pt,C_2,H_7,N_2]^+ + AuH \qquad (7.14)$$

$$[Pt,Au,C,H,N]^+ + CH_3NH_2 \quad \rightarrow \quad [Pt,C_2,H_5,N_2]^+ + AuH \qquad (7.15)$$

7.5 Comparison with the Reactivity of Bimetallic Catalysts

The reactivities of the Pt$_m$Au$_n$$^+$ ions may also be compared with those of related bimetallic clusters in the condensed phase. For instance, Aubart *et al.* reported complexes such as [Pt(AuPPh$_3$)$_8$](NO$_3$)$_2$ to efficiently catalyze H$_2$/D$_2$ equilibration.[209,210] Similar to the present findings, homonuclear gold clusters are unreactive. However, in contrast to the situation for the Pt$_m$Au$_n$$^+$ ions, the reactivities of the phosphine-stabilized clusters (with respect to H$_2$ activation) increase as a function of gold content.[210] Although it might well be the case that the activities of bimetallic platinum-gold clusters differ for reactions with CH$_4$ and O$_2$ on one hand and H$_2$ on the other, it also appears possible that the metal clusters' intrinsic reactivities in the H$_2$/D$_2$ equilibration reactions are shielded by ligand effects. Aubart *et al.* indicate that a weaker binding of the phosphine ligands to gold atoms might facilitate their dissociation which is essential for catalyst activation.[210] In fact, a theoretical investigation of H$_2$ dissociation on bare di- and trinuclear platinum-gold clusters found gold to decrease the clusters' reactivity,[211] in agreement with the present results. Moreover, a related study of the reactions of ethylene with Pt$_2$Au and PtAu$_2$ predicted a strong weakening of the metal-substrate interaction upon substitution of platinum by gold such that chemisorption no longer occurred.[212] Thus, the reactivities of gaseous bimetallic platinum-gold clusters toward different substrates are consistently diminished as a function of gold content. This trend corresponds to the behavior of bulk catalysts where gold is generally known to decrease catalytic activities.[55]

In conclusion, Jacobsen's interpolation scheme for predicting the reactivity of bimetallic surface catalysts can be successfully applied to rationalize the behavior of heteronuclear Pt$_m$Au$_n$$^+$ clusters in the gas phase. Apparently, the intermediate reactivity of the bimetallic systems constitutes a truly molecular phenomenon that is only relatively weakly disturbed by size-specific electronic effects. Consequently, a further investigation of the potential of Pt$_m$Au$_n$$^+$ clusters with respect to C–N coupling appears quite promising.

8 C–N Coupling by Bimetallic $Pt_mAu_n^+$ Clusters

The above findings reveal a lower gross reactivity of heteronuclear $Pt_mAu_n^+$ clusters in comparison to the homonuclear Pt_m^+ ions. With respect to the activation of methane, the electronically more saturated $Pt_mAu_n^+$ clusters bind the carbene fragment less strongly than their Pt_{m+n}^+ counterparts; this reduced interaction might facilitate the coupling of the activated first component, $i.e.$, CH_2, with a second substrate such as ammonia. Moreover, there is also direct experimental evidence for the positive effect of gold in terms of C–N coupling. Thus, among several mononuclear transition-metal carbenes MCH_2^+ investigated, $AuCH_2^+$ exhibited an enhanced tendency towards coupling with NH_3, reaction 8.1.[40] Remember, however, that $AuCH_2^+$ is not formed spontaneously upon reaction of ground-state Au^+ with CH_4 at ambient conditions.

$$AuCH_2^+ + NH_3 \quad \rightarrow \quad CH_2NH_2^+ + AuH \qquad (8.1)$$

In the following, the reactions of $Pt_mAu_nCH_2^+$ clusters, $m + n \leq 4$, with NH_3 are studied in order to probe the potential of these clusters with respect to C–N bond coupling. The experiments do not only include the Pt-rich carbene species directly accessible from CH_4 dehydrogenation but also take into account some less stable Au-rich carbene clusters that have to be prepared via alternative routes.

8.1 Reactions of $PtAuCH_2^+$ with NH_3

Upon reaction with NH_3, the heteronuclear carbene cluster $PtAuCH_2^+$ loses H_2, reaction 8.2 (Table 8.1). Formally similar reactions occur for $PtCH_2^+$ as well as for the corresponding $Pt_mCH_2^+$ clusters but only in the former case result in C–N bond formation, see Section 5.1. The ionic product $[Pt,Au,C,H_3,N]^+$ then undergoes consecutive reactions with a second NH_3 molecule, reactions 8.3 and 8.4 with an overall rate constant $k = 1.3 \times 10^{-10}$ $cm^3 s^{-1}$ ($\varphi = 0.066$) and a branching ratio 45 : 55.

$$PtAuCH_2^+ + NH_3 \qquad \rightarrow \qquad [Pt,Au,C,H_3,N]^+ + H_2 \qquad\qquad (8.2)$$

$$[Pt,Au,C,H_3,N]^+ + NH_3 \qquad \rightarrow \qquad [Pt,Au,C,H_4,N_2]^+ + H_2 \qquad\qquad (8.3)$$

$$\rightarrow \qquad [Pt,C,H_5,N_2]^+ + AuH \qquad\qquad (8.4)$$

The dehydrogenation observed in reaction 8.3 implies an activation of NH_3 that is inconsistent with $[Pt,Au,C,H_3,N]^+$ being a simple adduct of a $PtAuC^+$ entity and molecular NH_3. Instead, such a complex should only undergo degenerate ligand exchange and, if termolecular stabilization is feasible, association with a second NH_3 molecule, as $Pt_2C(NH_3)^+$ does, see Section 5.1. Behavior analogous to reaction 8.3, however, is observed for the aminocarbene complex $PtC(H)NH_2^+$ formed from atomic Pt^+ in the presence of CH_4 and NH_3.[39,40] This finding is a first indication that the reactivity of $PtAu^+$ resembles that of Pt^+ rather than of Pt_2^+ or the larger homonuclear platinum clusters. Similarly, the second product channel, reaction 8.4, brings about degradation of the metal core by loss of neutral AuH and, thus, does not agree with a $PtAuC(NH_3)^+$ structure, either. Interestingly, an analogous AuH elimination occurs as consecutive reaction with CH_3NH_2, reaction 7.14. Because the respective primary product $[Pt,Au,C,N,H_3]^+$ in the $PtAu^+/CH_3NH_2$ system has the same total formula like that arising from the $PtAuCH_2^+/NH_3$ combination, the remarkable parallel in reactivity of both species might point to their structural identity.

Further clues evolve from a comparison with the extensively studied Pt^+/CH_3NH_2 system. Here, labeling experiments yielded evidence for a selective 1,1-elimination of H_2 from the methyl group in CH_3NH_2.[39] This behavior is consistent with the high reactivity of Pt^+ toward CH_4 and its failure to dehydrogenate NH_3. As $PtAu^+$ also discriminates between CH_4 and NH_3, this ion is supposed to selectively attack the methyl group in CH_3NH_2 as well; the labeling experiments fully support this assignment (see Section 7.4). The resulting aminocarbene structure $PtAuC(H)NH_2^+$ is therefore assumed as a common intermediate of both the $PtAuCH_2^+/NH_3$ and $PtAu^+/CH_3NH_2$ routes.

Besides single dehydrogenation leading to $[Pt,Au,C,N,H_3]^+$, the reaction of $PtAu^+$ with CH_3NH_2 predominantly affords elimination of two molecules H_2, reaction 7.10 with $m = n = 1$. One possible explanation why the reaction of $PtAuCH_2^+$ with NH_3 does not yield this product relies on energetic arguments. Taking into account that $D_{298}(M^+–CH_2) > 465$ kJ mol^{-1} must hold for spontaneous dehydrogenation of CH_4, the $MCH_2^+ + NH_3$ channel lies at least 120 kJ mol^{-1} lower in energy than the M$^+$ + CH$_3$NH$_2$ asymptote.[180] Consequently, the approach of CH$_3$NH$_2$ to M$^+$ should liberate substantially more energy than complexation of MCH$_2^+$ by NH$_3$; this extra energy is available for surpassing putative barriers associated with double elimination of H$_2$ from [M,C,H$_5$,N]$^+$. Provided the validity of this hypothesis in the case of M = PtAu, external energy input to [Pt,Au,C,H$_3$,N]$^+$ originating from PtAuCH$_2^+$/NH$_3$ should induce further dehydrogenation. Indeed, CID of [Pt,Au,C,H$_3$,N]$^+$ produces the fragments PtAuC$^+$ and [Pt,Au,C,H,N]$^+$ in a ratio of about 2:1 for a range of collision energies, reactions 8.5 and 8.6 (at higher energies, atomic Au$^+$ and Pt$^+$ are generated as well). Thus, the expected dehydrogenation does occur lending strong support to an aminocarbene structure of the precursor ion. However, partial presence of a PtAuC(NH$_3$)$^+$ species fragmenting solely according to reaction 8.5 cannot be excluded.

$$[\text{Pt,Au,C,H}_3\text{,N}]^+ \quad \rightarrow \quad \text{PtAuC}^+ + \text{NH}_3 \qquad (8.5)$$

$$\rightarrow \quad [\text{Pt,Au,C,H,N}]^+ + \text{H}_2 \qquad (8.6)$$

Further mechanistic insight is achieved by labeling. Reaction of PtAuCD$_2^+$ with NH$_3$ gives a mixture of [Pt,Au,C,H$_3$,N]$^+$, [Pt,Au,C,H$_2$,D,N]$^+$, and [Pt,Au,C,H,D$_2$,N]$^+$ in a ratio of 70 ($\pm$ 15) : 100 : 80 ($\pm$ 15), reactions 8.2a - 8.2c.

$$\text{PtAuCD}_2^+ + \text{NH}_3 \quad \rightarrow \quad [\text{Pt,Au,C,H}_3\text{,N}]^+ + \text{D}_2 \qquad (8.2a)$$

$$\rightarrow \quad [\text{Pt,Au,C,H}_2\text{,D,N}]^+ + \text{HD} \qquad (8.2b)$$

$$\rightarrow \quad [\text{Pt,Au,C,H,D}_2\text{,N}]^+ + \text{H}_2 \qquad (8.2c)$$

Reactions 8.2b and c clearly demonstrate activation of NH$_3$. Once again, PtAu$^+$ behaves differently from Pt$_2^+$ but similarly to Pt$^+$. In the case of the latter, only the analogue of HD elimination 8.2b takes place, however,[39] whereas also losses of D$_2$ and H$_2$ occur in the reaction of PtAuCD$_2^+$ and NH$_3$. Reactions 8.2a and c point to an equilibration of all H and D atoms in the course of the dehydrogenation. Neglecting KIEs, a ratio of 17 : 100 : 50 is expected for reactions 8.2a - c with a completely statistical distribution. Experiment finds a much higher fraction of 8.2a which may be rationalized by some contribution of simple D$_2$/NH$_3$ exchange exclusively leading to PtAuC(NH$_3$)$^+$; the NH$_3$ loss observed upon CID (reaction 8.5) can also be considered as indication of such a process. This particular product channel might therefore correspond to the reactivity of Pt$_2$CD$_2^+$. The higher extent of reaction 8.2c than predicted by statistics can be ascribed to the operation of KIEs which are supposed to favor H_2 elimination in reaction 8.2c compared to losses of HD or D_2.

In summary, all experimental results agree that PtAu$^+$, unlike Pt$_2^+$, affords coupling of CH_4 and NH_3 (though perhaps not completely). Accordingly, the controlled weakening of the metal-carbene interaction in the heteronuclear cluster results in the desired effect and activates the CH_2 moiety for further reactions. The reaction pathway of the bimetallic system strongly resembles that inferred for PtCH$_2^+$/NH$_3$ where the final aminocarbene complex could be characterized also by quantum chemical methods.[40] This similarity is not necessarily expected *a priori* because the addition of an Au atom with its $5d^{10}6s^1$ valence-electron configuration is supposed to severely perturb the electronic situation of Pt$^+$. One borderline case would be the complete transfer of gold's single $6s$ electron to Pt. Although this charge distribution is disfavored by the metals' *IE*s and thus unlikely for bare PtAu$^+$ (*ab initio* calculations predict an equal partitioning of the positive charge),[213] it appears more probable for the PtAuCH$_2^+$ species because the resulting $5d^96s^1$ configuration of Pt would be well suited to form a double bond with the CH_2 fragment. The PtCH$_2$ entity then could coordinate to the Au$^+$ atom yielding a bridged structure (Scheme 8.1) and stabilize the gold cation by the transfer of electron density. A similar situation is found in the complexes of Au$^+$ with other electron-donating ligands, see Section 7.3.[151,199]

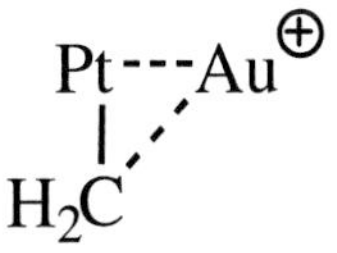

$$\text{Pt} \text{---} \text{Au}^{\oplus}$$
$$| \quad \diagup$$
$$\text{H}_2\text{C}$$

Scheme 8.1.

Additionally, the Au atom is likely to actively participate in the coupling reaction of PtAuCH$_2^+$ with NH_3. In analogy to the theoretical results obtained for the reaction of mononuclear PtCH$_2^+$ with NH_3,[40] strong metal-hydrogen interactions are also assumed in the PtAuCH$_2^+$/NH$_3$ system to account for the occurrence of hydrogen rearrangements as revealed by the labeling experiments. Because of atomic gold's high affinity for H (D_0(Au–H) = 300 kJ mol^{-1})[165] its involvement in these H shifts appears probable. The consecutive loss of AuH, reaction 8.4, can be regarded as direct evidence for such processes.

Before addressing the behavior of the larger bimetallic clusters, the reactivity of $Au_2CH_2^+$ is probed as well. Whereas $Au_2CH_2^+$ cannot be produced by reaction of the bare metal cluster with CH_4, application of CH_3Cl or CH_3Br proves successful in this respect, reaction 8.7.

$$Au_2^+ + CH_3X \rightarrow \quad Au_2CH_2^+ + HX, \ X = Cl, Br \quad\quad\quad (8.7)$$

Except of a very low tendency towards adduct formation, $Au_2CH_2^+$ does not react with NH_3 ($\varphi \leq 5 \times 10^{-3}$). This finding is somewhat surprising because the change from $PtAuCH_2^+$ to $Au_2CH_2^+$ is expected to further decrease the metal-carbene binding energy and, thus, to enhance the system's reactivity with respect to C–N coupling. While the latter reaction obviously does not take place with an appreciable efficiency, the formation of carbide complexes as observed in the case of $Pt_2CH_2^+$ does not occur either for $Au_2CH_2^+$. As mentioned above, such a process is believed to require a strong binding between the carbon atom and the metal core.

Table 8.1. Bimolecular rate constants k and efficiencies φ for the reactions of $Pt_mAu_nCH_2^+$ clusters with NH_3.

products	m	n	k / 10^{-10} cm^3 s^{-1}	φ
$CH_2NH_2^+ + Pt_mAu_nH$	1	0	4.3	0.21
	0	1	12.0	0.60
$[Pt_m,Au_n,C,H_3,N]^+ + H_2$	1	0	1.6	0.09
	2	0	9.7	0.49
	1	1	6.0	0.30
	0	2	< 0.1	< 0.01
	3	0	9.6	0.48
	2	1	8.3	0.42
	1	2	2.2	0.11
	4	0	17.0	0.86
	3	1	11.0[a]	0.54
	2	2		
	1	3	0.3	0.02
$NH_4^+ + Pt_mAu_nCH^+$	1	0	0.3	0.02
$[Pt_m,Au_{n-1},C,H_5,N]^+ + Au$	1	2	5.1	0.25

[a] Combined value for $Pt_3AuCH_2^+$ and $Pt_2Au_2CH_2^+$, see Section 7.1 for details.

8.2   Reactions of the Larger $Pt_mAu_nCH_2^+$ Clusters with NH_3

Similar to $Pt_3CH_2^+$, the Pt-rich bimetallic cluster $Pt_2AuCH_2^+$ efficiently loses H_2 upon exposure to NH_3 (Table 8.1), reaction 8.8 with $m = 2$ and $n = 1$.

$$Pt_mAu_nCH_2^+ + NH_3 \quad \rightarrow \quad [Pt_m,Au_n,C,H_3,N]^+ + H_2 \qquad (8.8)$$

As demonstrated above, this observation alone does not suffice to distinguish between the two different reactivity patterns identified so far, *i.e.*, carbide generation *versus* C–N coupling. A first indication about the structure of the ion produced is provided by consideration of consecutive reactions. Unlike the aminocarbenes derived from Pt^+ and $PtAu^+$, $[Pt_2,Au,C,H_3,N]^+$ does not activate a second NH_3 equivalent, but simply adds a further NH_3 molecule. This type of reactivity corresponds to that observed for the homonuclear carbide clusters $Pt_mC(NH_3)^+$, such that an analogous $Pt_2AuC(NH_3)^+$ structure appears likely in the case of the heteronuclear cluster as well. This assignment is further supported by CID experiments of $[Pt_2,Au,C,H_3,N]^+$ that result in facile loss of NH_3 without any indication of its activation. Besides simple NH_3 elimination, reaction 8.9, additional loss of an Au atom occurs, reaction 8.10, where the neutral species might either correspond to $Au(NH_3)$ or to bare Au concomitant with NH_3. Given the appreciable binding energy of neutral $Au(NH_3)$, $D_e(Au–NH_3) = 76 \pm 6$ kJ mol^{-1} according to *ab initio* calculations,[198] the formation of the bound complex appears somewhat more probable, although it is entropically disfavored in comparison to the separated species. The replacement of a metal atom by NH_3 does not have a counterpart in the case of the homonuclear Pt clusters and thus points to the weaker binding of the gold atom in the cluster.

$$[Pt_2,Au,C,H_3,N]^+ \quad \rightarrow \quad Pt_2AuC^+ + NH_3 \qquad (8.9)$$

$$[Pt_2,Au,C,H_3,N]^+ \quad \rightarrow \quad Pt_2C^+ + [Au,N,H_3] \qquad (8.10)$$

Further evidence for the formation of a carbide species is provided by labeling experiments. Upon reaction with NH_3, $Pt_2AuCD_2^+$ exclusively loses D_2, reaction 8.8a with $m = 2$ and $n = 1$, thereby ruling out the occurrence of N–H bond activation required for C–N coupling.

$$Pt_mAu_nCD_2^+ + NH_3 \quad \rightarrow \quad [Pt_m,Au_n,C,H_3,N]^+ + D_2 \qquad (8.8a)$$

In the case of the Au-rich trinuclear system $PtAu_2^+$, the carbene cluster can no longer be generated by dehydrogenation of CH_4. Instead, like for the atomic Au^+ ion[214] and dinuclear

Au_2^+ (see Section 8.1), treatment with CH_3Cl or CH_3Br affords the desired carbene complex $PtAu_2CH_2^+$, reaction 8.11 with $m = 1$ and $n = 2$.

$$Pt_mAu_n^+ + CH_3X \quad \rightarrow \quad Pt_mAu_nCH_2^+ + HX, \; X = Cl, Br \qquad (8.11)$$

Because Pt_2Au^+ undergoes the same reaction, however, a mass separation of the two overlapping distributions of cluster ions is impossible. Based on the isotopomers' predicted abundances, the $m/z = 604$ ion is mass-selected prior to exposure to NH_3 in order to maximize the $PtAu_2^+/Pt_2Au^+$ ratio. The actual composition of this peak can be probed by reaction with CH_4 that discriminates between both species; a constant $PtAu_2^+/Pt_2Au^+$ ratio of 60:40 ($\pm$ 5 %) is found throughout the experiment.

Treatment of the $m/z = 604$ ion with NH_3 brings about losses of H_2, reaction 8.8 with $m = 1$ and $n = 2$, and Au, reaction 8.12. As the latter process does not occur for $Pt_2AuCH_2^+$ it can be unambiguously attributed to $PtAu_2CH_2^+$. Taking the $PtAu_2^+/Pt_2Au^+$ ratio determined into account, an efficiency $\varphi = 0.36$ and branching ratios of 30 and 70 % for H_2 and Au elimination, respectively, can thus be derived for pure $PtAu_2^+$.

$$PtAu_2CH_2^+ + NH_3 \quad \rightarrow \quad [Pt,Au,C,H_5,N]^+ + Au \qquad (8.12)$$

Reaction 8.12 can be viewed as a formal substitution of one Au atom by NH_3 which does not involve the carbene fragment and therefore is irrelevant as far as C–N bond formation is concerned. Interestingly, the bare $PtAu_2^+$ ion does not undergo an analogous degradation reaction in the presence of NH_3, see Section 7.4. Apparently, the carbene ligand destabilizes the metal core such that the substitution of an Au atom by NH_3 becomes feasible. Note that a similar ligand effect is observed for $Au_3^+/Au_3NH_3^+$, where the bare cluster is inert as well while the ligated system undergoes fragmentation (see Section 7.4).

Regarding dehydrogenation (reaction 8.8 with $m = 1$ and $n = 2$), the most interesting question again is whether or not this process involves activation of NH_3. With $PtAu_2CD_2^+$, only D_2 is lost upon interaction with NH_3, reaction 8.8a with $m = 1$ and $n = 2$, thus strongly suggesting a carbide structure for the product ion. Accordingly, $PtAu_2CH_2^+$ behaves similarly to $Pt_2AuCH_2^+$ and $Pt_3CH_2^+$, but differs from $PtAuCH_2^+$. However, the predominance of Au loss compared to dehydrogenation reflects a lowered tendency towards carbide formation for $PtAu_2CH_2^+$, which is in agreement with the expected weaker interaction of the Au-rich metal core and carbon. Despite this weaker binding, the carbene fragment in $PtAu_2CH_2^+$ is obviously not available for coupling with NH_3, thus resembling the situation encountered for Au$_2$CH$_2^+$. While a comparison with the reactivity of the trinuclear Au$_3$CH$_2^+$ cluster would be desirable, unfortunately no practicable gas-phase synthesis for this ion is found as the reactions of Au$_3^+$ with CH$_4$, CH$_3$Cl, CH$_3$Br, and CH$_3$I, respectively, do not yield the desired Au$_3$CH$_2^+$ ion.

The reactions of the tetranuclear clusters do not show significantly new features. Pt$_3$AuCH$_2^+$ and Pt$_2$Au$_2$CH$_2^+$ (see Section 7.1) efficiently undergo dehydrogenation in their reactions with NH$_3$. Labeling experiments again indicate the formation of carbide complexes, reaction 8.8a. The same holds true for the reaction of PtAu$_3$CH$_2^+$, although this species displays a considerably decreased reaction efficiency compared to its Pt-rich counterparts (Table 8.1). Presumably, the lower rate constant derived for the dehydrogenation of PtAu$_3$CH$_2^+$ reflects a less favorable thermochemistry in terms of a weaker interaction between the carbon atom and the electronically more saturated Au-rich metal core. In contrast to its smaller congener PtAu$_2$CH$_2^+$, cluster degradation analogous to reaction 8.12 does not occur for PtAu$_3$CH$_2^+$, thus pointing to distinct stabilities of the different clusters.

In conclusion, the larger bimetallic Pt$_m$Au$_n$CH$_2^+$ clusters, $m + n \geq 3$, behave similarly in that they do not mediate C–N bond formation in the presence of NH$_3$ but preferentially yield carbide complexes. A closer inspection reveals an important difference between the reactivities of the Pt-rich and the Au-rich carbenes, however. Whereas the former, like the homonuclear Pt$_m$CH$_2^+$ clusters, are efficiently transformed to carbide species in the presence of NH$_3$, the tendency towards dehydrogenation is strongly decreased for the latter (Table 8.1). The experimental results suggest that the presence of at least two Pt atoms in the cluster is necessary for facile carbide generation. This finding nicely agrees with the situation of the homonuclear carbenes Pt$_m$CH$_2^+$ whose reactions with NH$_3$ exclusively yield the carbide species for $m \geq 2$ (see section 5.1). Most probably, the stronger interaction of the carbon atom with Pt$_m^+$ and Pt$_m$Au$^+$, $m \geq 2$, requires a multi-center bonding situation.

With regard to the intended C–N coupling, a dilemma evolves from the observed trends in reactivity. Whereas a high platinum content on one hand is needed to dehydrogenate CH$_4$ in the first reaction step, the presence of two or more Pt atoms in the cluster on the other results in an unfavorably strong metal-carbon interaction. As a consequence of this stronger binding, the Pt-rich carbenes are no longer activated for coupling with NH$_3$. These opposing demands for the overall process of C–N bond formation are only fulfilled for Pt^+ and $PtAu^+$ that apparently meet the requirement of an especially well-balanced metal-carbene interaction. Accordingly, the metal-carbene binding energy is considered as the prime factor controlling the clusters' reactivities toward NH_3, much alike the situation in Co-Mo-catalyzed ammonia synthesis where the metal-nitride interaction was proven to be crucial.[187] In contrast to the heterogeneous Co-Mo system, however, other effects appear to be important as well.[215] Particularly, the lacking ability of $Au_2CH_2^+$, $PtAu_2CH_2^+$, and $PtAu_3CH_2^+$ to mediate C–N coupling points to the operation of such specific electronic or structural effects because these relatively weakly bound carbenes are expected to be reactive from a merely energetic point of view. Clearly, this finding calls for further efforts aiming at a better understanding of the factors that distinguish $PtAu^+$ from all other bimetallic $Pt_mAu_n^+$ cluster ions.

9   Reactivities of PtCu⁺ and PtAg⁺

Among the different factors that potentially contribute to the enhanced reactivity of PtAu⁺ in terms of C–N coupling, the electronic configuration might be of particular importance. Size-specific electronic effects are well-known to strongly influence the reactivity of metal clusters,[200] and several examples indicative of such behavior have been encountered in the course of the present work. A possible means to assess the role of electronic properties consists in a comparison of isoelectronic species and their reactivities. In the case of PtAu⁺, a comparison with PtCu⁺ and PtAg⁺ appears particularly interesting because these ions are supposed to closely resemble the valence-electron configuration of PtAu⁺, while presumably significantly differing in electron affinities, orbital sizes, bond lengths, etc. Similarity or dissimilarity in reactivity should reveal which of these factors matter most.

Specifically, the reactions with CH_4, O_2, and NH_3 are investigated; for comparison, the reactivities of Cu_2^+ and Ag_2^+ are included as well. Particular attention is paid to the reactions of the metal carbenes with NH_3.

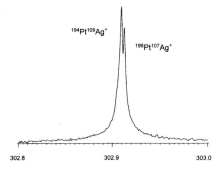

Figure 9.1. High-resolution mass spectrum of PtAg⁺ (mass range m/z = 302.8 - 303.0). The relative mass difference between ¹⁹⁴Pt¹⁰⁹Ag⁺ and ¹⁹⁶Pt¹⁰⁷Ag⁺ is $\Delta m/m \approx 8$ ppm. The absolute masses are slightly shifted upwards because of loss in magnetic flux density B_z since the last calibration.

Like PtAu$^+$, the heteronuclear clusters PtM$^+$, M = Cu, Ag, are produced by laser vaporization/ionization of the corresponding solid alloys (2:1 weight ratio in both cases). Pure metal targets are used to generate Cu$_2$$^+$ and Ag$_2$$^+$. For the reactivity studies, in each case the most abundant dinuclear ion is selected, *i.e.*, ^{63}Cu$_2$$^+$, ^{107}Ag^{109}Ag$^+$, ^{63}Cu^{196}Pt$^+$/^{65}Cu^{194}Pt$^+$, and ^{107}Ag^{196}Pt$^+$/^{109}Ag^{194}Pt$^+$, respectively; the latter two combinations of isotopomers are only resolved in the single-ion detection-mode (Figure 9.1).

9.1   Reactions of the Bare PtM$^+$ Ions

Similarly to Pt$_2$$^+$ and PtAu$^+$, both PtCu$^+$ and PtAg$^+$ efficiently activate CH$_4$ (Table 9.1), reaction 9.1.

$$\text{PtM}^+ + \text{CH}_4 \quad \rightarrow \quad \text{PtMCH}_2^+ + \text{H}_2 \hspace{3cm} (9.1)$$

As the homonuclear coinage-metal dimers M$_2$$^+$ (M = Cu, Ag, Au) prove to be inert,[216] the Pt center appears essential for the reaction. This finding meets expectations because a sufficient thermochemical stability of the metal carbene formed is a prerequisite for the spontaneous dehydrogenation of CH$_4$. Such a strong M$_2$$^+$–CH$_2$ interaction requires a double bond and, thus, at least two free valences at the metal core that are not available in the cases of Cu$_2$$^+$, Ag$_2$$^+$, and Au$_2$$^+$. Because promotion energies are rather high for the first- and second-row transition metals,[33] *ds* hybridization is not a viable option for Cu$_2$$^+$ and Ag$_2$$^+$ either.

Exposure of PtCu$^+$ and PtAg$^+$ to O$_2$ results in cluster degradation, reaction 9.2. Platinum's affinity for the +IV oxidation state encountered in PtO$_2$ is also well-known from classical solution chemistry; in contrast, copper and silver prefer lower oxidation states, in accordance with the product distribution observed.[130] The bias of copper and silver against high oxidation states also accounts for the lacking reactivity of Cu$_2$$^+$ and Ag$_2$$^+$ toward O$_2$.[216,217]

$$\text{PtM}^+ + \text{O}_2 \hspace{2cm} \rightarrow \hspace{1cm} \text{M}^+ + \text{PtO}_2 \hspace{3cm} (9.2)$$

The efficiency of cluster degradation decreases when changing M from Cu to Ag, and the reaction completely ceases for Au (Table 9.1). Presumably, the high ionization energy of gold, *IE*(Au) = 9.23 eV, strongly disfavors the generation of Au$^+$ in comparison to Cu$^+$ and Ag$^+$ (*IE*(Cu) = 7.73 and *IE*(Ag) = 7.58 eV).[118] Moreover, the similar sizes and energies of the orbitals of Pt and Au are supposed to lead to a better overlap and a stronger binding that further hinders cluster degradation for PtAu⁺.

Table 9.1. Bimolecular rate constants k and efficiencies φ for the reactions of the dinuclear ions PtCu⁺, PtAg⁺, and PtAu⁺ as well as the corresponding carbene species.

reaction		$k\,/\,10^{-10}\,\mathrm{cm^3\,s^{-1}}\,(\varphi)$		
		M = Cu	Ag	Au
PtM⁺ + CH₄	→ PtMCH₂⁺ + H₂	5.8	6.0	6.4
		(0.59)	(0.63)	(0.67)
PtM⁺ + O₂	→ M⁺ + PtO₂	0.35	0.04	0
		(0.064)	(0.01)	
PtM⁺ + NH₃	→ MNH₃⁺ + Pt	8.5	3.9	0
		(0.42)	(0.19)	
	→ PtNH₃⁺ + M	0	0	2.8
				(0.14)
PtMCH₂⁺ + NH₃	→ [Pt,M,C,H₃,N]⁺ + H₂	5.2	4.9	6.0
		(0.26)	(0.24)	(0.30)

A different kind of cluster degradation occurs upon reaction of PtCu⁺ and PtAg⁺ with NH₃, reaction 9.3. In this reaction type, the basic substrate replaces one metal atom and yields MNH₃⁺ complexes. Whereas neutral Pt is lost from PtCu⁺ and PtAg⁺, the analogous reaction of PtAu⁺ gives PtNH₃⁺ and Au (reaction 7.7). Similar degradation processes are observed for the homonuclear M₂⁺ coinage-metal clusters.[216]

$$PtM^+ + NH_3 \quad \rightarrow \quad MNH_3^+ + Pt \tag{9.3}$$

The product distributions found for the heteronuclear clusters can be explained by simple thermochemical considerations. An analysis of the heats of formation for the two alternative product combinations, MNH₃⁺ + Pt *versus* PtNH₃⁺ + M, demonstrates that the former exit channel clearly lies lower in energy for M = Cu and Ag (Table 9.2), in agreement with experiment. For M = Au, the larger uncertainties of the thermochemical data available do not allow an unambiguous assessment. Notwithstanding, the relative stability of the PtNH₃⁺/M product channel is significantly enhanced in this case compared to the situation for M = Cu and Ag, in agreement with the different reactivities observed.

Table 9.2. Heats of formation $\Delta_f H_0°$ (in kJ mol^{-1}) for the two alternative product channels observed in the reactions of PtM$^+$ with NH$_3$, MNH$_3$$^+$ + Pt *versus* PtNH$_3$$^+$ + M.[a]

M	$\Delta_f H_0°$(MNH$_3$$^+$ + Pt)	$\Delta_f H_0°$(PtNH$_3$$^+$ + M)	$\Delta\Delta_f H_0°$
Cu	1366 ± 15	1451 ± 12	85 ± 19
Ag	1365 ± 13	1398 ± 12	33 ± 18
Au	1479 ± 30	1479 ± 12	0 ± 32

[a] Calculated from D_0(M$^+$–NH$_3$) (taken from ref. [113,197,218,219]), $\Delta_f H_0°$(M), $\Delta_f H_0°$(NH$_3$) (taken from ref. [180]) and *IE*(M) (ref. [118]).

9.2 Reactions of the PtMCH$_2$$^+$ Ions with NH$_3$

The reactivity of the heteronuclear clusters toward NH$_3$ changes completely in the presence of a CH$_2$ fragment at the metal core. Instead of cluster degradation, exclusive dehydrogenation takes place (Table 9.1), reaction 9.4.

$$\text{PtMCH}_2{}^+ + \text{NH}_3 \qquad \rightarrow \qquad [\text{Pt,M,C,H}_3,\text{N}]^+ + \text{H}_2 \qquad (9.4)$$

$$[\text{Pt,M,C,H}_3,\text{N}]^+ + \text{NH}_3 \qquad \rightarrow \qquad [\text{Pt,M,C,H}_4,\text{N}_2]^+ + \text{H}_2 \qquad (9.5)$$

For both M = Cu and M = Ag, the primary product undergoes a consecutive dehydrogenation, reaction 9.5, in the former case with rather high (φ = 0.23) and in the latter with low ($\varphi \approx 0.01$) efficiency; an analogous reaction also takes place for M = Au (φ = 0.03), see Section 8.1. The fact that the primary products [Pt,M,C,H$_3$,N]$^+$ still are capable of NH$_3$ activation strongly suggests that already the first dehydrogenation involves activation of NH$_3$. This issue can be further probed by isotopic labeling.

The reaction of PtAuCD$_2$$^+$ with NH$_3$ leads to the elimination of D$_2$, HD, and H$_2$ in a ratio of 70 (± 15): 100 : 80 (± 15), see Section 8.1. Again, quite a similar behavior is found for PtCuCD$_2$$^+$ and PtAgCD$_2$$^+$. For the former, the ratio for losses of D$_2$, HD, and H$_2$ amounts to 15 (± 10): 100 : 25 (± 20) whereas it corresponds to ≤ 5 : 100 : 30 (± 5) in the latter case. The rather high ranges of uncertainty assigned particularly to the product distribution for M = Cu mainly result from the complicating effects of D/H exchange reactions for the deuterium containing product ions, which require a determination of the branching ratios by extrapolation to zero reaction time.

The experimentally observed product ratios can be compared with two borderline cases. Neglecting KIEs, a purely statistical H/D equilibration should yield a 17 : 100 : 50 distribution for losses of D$_2$, HD, and H$_2$, respectively. PtCuCD$_2^+$ and PtAgCD$_2^+$ behave accordingly in that they yield significantly larger fractions of H$_2$ than of D$_2$ elimination. In contrast, both product channels almost have equal probability for PtAuCD$_2^+$, which possibly indicates the presence of a second pathway leading to direct D$_2$/NH$_3$ exchange without NH$_3$ activation in this case, see Section 8.1. Both PtCuCD$_2^+$ and PtAgCD$_2^+$ exhibit an increased fraction of HD loss compared to statistical scrambling. This behavior resembles the reactivity of mononuclear PtCD$_2^+$ that exclusively eliminates HD upon reaction with NH$_3$ affording PtC(D)NH$_2^+$.[39,40] Given the similarity between the reactivities of PtCD$_2^+$ and PtMCD$_2^+$, M = Cu and Ag, toward NH$_3$, analogous structures of the product ions appear likely. The partial H/D scrambling found for the dinuclear ions might point to a secondary interaction between the M atom and the attacking NH$_3$ molecule in the course of the reaction. In the case of PtAuCD$_2^+$, the enhanced scrambling observed is in line with a particularly favorable M–H binding for Au, which results from both the relatively high electronegativity of gold and the large spatial extension of its 6s orbital.

Further evidence for the formation of aminocarbene complexes is provided by the consecutive reactions of the labeled ions. In the case of mononuclear PtC(D)NH$_2^+$, exclusive loss of HD occurs in the reaction with NH$_3$, which is consistent with the theoretically predicted bisaminocarbene structure of the secondary product.[39,40] Whereas H/D exchange reactions presumably compete with the inefficient consecutive reactions of the [Pt,Ag,C,H$_x$,D$_y$,N]$^+$ species, $x + y = 3$, and thus prevent an assessment of the products' labeling distribution, the situation is more favorable for its faster reacting copper counterparts. Here, one finds the products [Pt,Cu,H$_4$,N$_2$]$^+$ and [Pt,Cu,H$_3$,D,N$_2$]$^+$ in a ratio of 100 : 20 ($\pm$ 10). This distribution fully agrees with the transformation of aminocarbene into bisaminocarbene structures under substitution of the carbene H or D atom, respectively, by an NH$_2$ group, reactions 9.5a - c.

$$\text{PtCuC(H)NH}_2^+ + \text{NH}_3 \quad \rightarrow \quad \text{PtCuC(NH}_2)_2^+ + \text{H}_2 \qquad (9.5a)$$

$$\text{PtCuC(D)NH}_2^+ + \text{NH}_3 \quad \rightarrow \quad \text{PtCuC(NH}_2)_2^+ + \text{HD} \qquad (9.5b)$$

$$\text{PtCuC(D)NHD}^+ + \text{NH}_3 \quad \rightarrow \quad \text{PtCuC(NH}_2)(\text{NHD})^+ + \text{HD} \qquad (9.5c)$$

In conclusion, both $PtCu^+$ and $PtAg^+$ display reactivities much resembling that of $PtAu^+$. Particularly, all three dinuclear platinum coinage-metal cluster ions PtM^+ mediate coupling of CH_4 and NH_3, whereas neither Pt_2^+ nor M_2^+ do so.[220] This finding strongly indicates that the very electronic structure of the PtM^+ ions is crucial for achieving C–N bond formation between CH_4 and NH_3. In contrast, the atomic radius, *IE*, and orbital energies of M appear to be much less important as the variation of these factors by going from Cu via Ag to Au hardly affect the ions' reactivity with regard to C–N coupling.

10   Conclusions

The present work explores the reactivities of gaseous platinum clusters toward a wide range of substrates. For all processes investigated, their potential usage as gas-phase models of the corresponding reactions involving heterogeneous platinum catalysts forms the main focus. The results can be considered from two different perspectives. First, the general performance of the applied approach is assessed and contrasted with that of alternative gas-phase models relying on mononuclear transition-metal ions. The second aspect concerns the potential relevance of the results obtained to technically important platinum catalysts.

Performance of the Model.   A comparison between the chemical behavior of the bare Pt_m^+ clusters and the reactivity of macroscopic platinum surfaces reveals remarkable parallels, thus encouraging the use of the cluster ions as gas-phase models for heterogeneous catalysts. Specifically, the reactions of the Pt_m^+ clusters resemble the behavior of platinum catalysts at elevated temperatures. This finding meets expectations because the complexation energy released upon the encounter of the ionic cluster and the neutral substrate is distributed over a relatively small number of internal degrees of freedom only, thereby significantly increasing the system's effective temperature.[221] In the case of mononuclear transition-metal ions, this effect may result in excessively high vibrational excitations which prevent comparisons with macroscopic systems under typical conditions. Similarly, the availability of an internal heat bath in the case of the clusters enhances their tendency towards association reactions in comparison to mononuclear systems. Given the importance of such association processes as elementary steps in heterogeneous catalysis, the use of platinum clusters as model compounds appears advantageous in this respect.

Another notable point concerns the metal's charge density. Enlargement of the metal core reduces the charge density of the mono-cationic clusters and thus asymptotically approaches electroneutrality of the solid state. However, this effect does not appear to influence the reactivity of the platinum clusters very strongly. Although the behavior of atomic Pt^+ differs from that of its larger homologues in several aspects, only the increased tendency towards hydride abstraction observed for the former can be attributed to its higher positive charge density. In line with the inferred relatively little influence of the effective charge, the reactions of anionic Pt_m^- clusters with CH_4 and N_2O closely resemble those of their cationic counterparts.[57,61] This finding further suggests that the presence of a full positive charge does not govern the reactivity of mononuclear transition-metal cations M^+ as much as one might assume.

In contrast, a major difference between atomic Pt^+ and its larger homologues evolves from the possibility of multi-center binding of a substrate in the case of the clusters. The interaction with more than a single metal center gives rise to stronger bonds and thus also allows the stabilization of highly unsaturated fragments X such as the carbene moiety or even a bare carbon atom. Consequently, these strongly bound fragments are no longer activated for coupling reactions such that the reactivity of Pt_mX^+ clusters is drastically diminished in comparison to their mononuclear counterpart PtX^+. In view of the well-known importance of bridging coordination modes for the interaction of numerous substrates with extended metal surfaces,[222] the inclusion of more than a single metal atom in the corresponding gas-phase models appears highly indicated. Note that these considerations should not apply to platinum only, but equally hold for other transition metals and their binding properties as well. The rather limited thermochemical data available for M_mX^+ species accordingly show a significant increase in the M_m–X^+ bond-dissociation energies when going from mono- to polynuclear clusters.[129,157,223]

Obviously, another advantage of the cluster model lies in the possibility to combine two different metals. Thereby, mimicking cooperative effects in general and bimetallic catalysis in particular should become feasible. The overall reactivity attenuation observed for the heteronuclear $Pt_mAu_n^+$ clusters upon increasing the gold content indeed parallels the poisoning effect of gold known for many cases in heterogeneous catalysis.

Clearly, the extension from mononuclear systems to small metal clusters significantly enhances the similarities with real catalysts and, thus, leads to a substantial improvement of the gas-phase model. The considerable efforts required for cluster generation and handling therefore certainly appear justified. Moreover, the experiments promise to yield even further insight when completed by theoretical studies. The example of mononuclear transition-metal ions and their reactivity demonstrates the benefit of this combined approach.[224] Whereas the accurate treatment of transition-metal clusters, particularly those consisting of $5d$ elements, seems hardly possible for current *ab-initio* techniques, the rapid progress in both computing power and method development is likely to bring about a change for the better in the near future.

Implications for Heterogeneous Platinum Catalysts. Among the results of the present thesis, the findings regarding methane activation and coupling with ammonia arguably are the most relevant from a practical point of view. The particular interest in this conversion is due to its industrial implementation in terms of the DEGUSSA process that employs platinum contacts for the large-scale synthesis of hydrogen cyanide from methane and ammonia. A better mechanistic understanding achieved for this process might eventually lead to catalyst refinement.

As shown in the present work, platinum clusters Pt_m^+ succeed in the activation of CH_4 but do not mediate coupling of the resulting carbene species $Pt_mCH_2^+$ with NH_3. The formation of $Pt_mC(NH_3)^+$ carbide complexes observed instead can be regarded as gas-phase analogue of the unwanted production of soot on the heterogeneous catalyst. For the gaseous platinum clusters, the total preponderance of carbide formation over C–N coupling can be explained by stabilization of the bare carbon atom by multi-center bonding. The occurrence of soot formation in the real heterogeneous process might indicate that the binding of carbon to the platinum surface is somewhat shifted from the optimum towards a too strong interaction as well.

A rational approach to promote the intended C–N coupling aims at weakening of the platinum-carbene binding energy. In the case of the gas-phase model, this modification is accomplished by the change from pure Pt_m^+ to mixed $Pt_mAu_n^+$ clusters. Note that the use of bimetallic cobalt-molybdenum catalysts suggested for ammonia synthesis pursues a conceptually similar strategy.[187] The smallest conceivable model of a bimetallic platinum-gold catalyst, *i.e.*, $PtAu^+$, indeed mediates C–N bond coupling of CH_4 and NH_3, probably yielding the aminocarbene complex $PtAuC(H)NH_2^+$ as product. This finding lends support to the assumption that controlled weakening of the metal-carbene binding activates the CH_2 fragment with respect to coupling reactions. Similarly, gold additives might enhance the performance of platinum catalysts in the DEGUSSA process.

The failure of the larger bimetallic $Pt_mAu_nCH_2^+$ clusters to afford C–N coupling shows, however, that consideration of the metal-carbene binding energy alone does not suffice for a complete understanding of the clusters' reactivities.[215] Rather, the distinct electronic situation in the dinuclear system seems to be crucial for the outcome of the reaction with ammonia as well. This assessment is supported by the occurrence of C–N coupling in the reactions of the related species $PtCuCH_2^+$ and $PtAgCH_2^+$. Furthermore, these findings also suggest possible positive effects of copper or silver additives on the activity of the DEGUSSA catalyst.

Outlook. With regard to the DEGUSSA process, a possible extension of the present work might probe some of the inferences from the gas-phase model under heterogeneous conditions. Here, the performance of bimetallic Pt/Cu, Pt/Ag, and Pt/Au catalysts appears to be a particularly interesting question. Ideally, experiments tackling this problem would also vary the microscopic structure of the catalysts' surfaces in a controlled manner. On the basis of the present results, one might expect a strong dependence of the catalytic activity on the degree of dispersion.

A second direction of future efforts may address the reactivities of further transition-metal clusters. Given that the heavy Pt_m^+ clusters with their broad isotope distributions can be handled by FT-ICR mass spectrometry, this technique should also allow the detailed investigation of other systems. A thus achieved broader data base is likely to reveal general trends in the chemical behavior of transition-metal clusters. For example, already the present findings point to the importance of bridging coordination modes in ligated clusters M_mX^+ and their influence on the systems' reactivities.

Besides homonuclear M_m^+ cluster ions, bimetallic clusters $M_mM'_n^+$ form an especially intriguing target. The present work demonstrates that laser vaporization/ionization of the corresponding alloys constitutes a fairly general access to these species. Like in the case of the $Pt_mAu_n^+$ clusters, the addition of a second metal might permit a selective manipulation of the system's reactivity.

With refined gas-phase models of solid-state catalysts within reach, mass spectrometry will provide an important impulse to heterogeneous catalysis in the near future, as it already does at present in the case of homogeneous catalysis.[225] Thus, gas-phase techniques will continue to contribute to the solution of the most urgent problems contemporary chemistry has to face.

11 References and Notes

[1] H. Arakawa, M. Aresta, J. N. Armor, M. A. Barteau, E. J. Beckman, A. T. Bell, J. E. Bercaw, C. Creutz, E. Dinjus, D. A. Dixon, K. Domen, D. L. DuBois, J. Eckert, E. Fujita, D. H. Gibson, W. A. Goddard, D. W. Goodman, J. Keller, G. J. Kubas, H. H. Kung, J. E. Lyons, L. E. Manzer, T. J. Marks, K. Morokuma, K. M. Nicholas, R. Periana, L. Que, J. Rostrup-Nielson, W. M. H. Sachtler, L. D. Schmidt, A. Sen, G. A. Somorjai, P. C. Stair, B. R. Stults, W. Tumas, *Chem. Rev.* **2001**, *101*, 953.

[2] M. Gerloch, E. C. Constable, *Transition Metal Chemistry*, VCH, Weinheim, **1994**.

[3] R. H. Crabtree, *J. Chem. Soc., Dalton Trans.* **2001**, 2437.

[4] J. A. Labinger, J. E. Bercaw, *Nature* **2002**, *417*, 507.

[5] C. Elschenbroich, *Organometallchemie*, 4th ed., Teubner, Stuttgart, **2002**.

[6] G. B. Kauffman, *Enantiomer* **1999**, *4*, 609.

[7] J. M. Thomas, *Angew. Chem.* **1994**, *106*, 963; *Angew. Chem. Int. Ed. Engl.* **1994**, *33*, 913.

[8] J. F. Kriz, T. D. Pope, M. Stanciulescu, J. Monnier, *Ind. Eng. Chem. Res.* **1998**, *37*, 4560.

[9] A. Hollo, J. Hancsok, D. Kallo, *Appl. Catal. A - Gen.* **2002**, *229*, 93.

[10] M. M. Bhasin, J. H. McCain, B. V. Vora, T. Imai, P. R. Pujado, *Appl. Catal. A - Gen.* **2001**, *221*, 397.

[11] P. Tetenyi, V. Galsan, *Appl. Catal. A - Gen.* **2002**, *229*, 181.

[12] A. W. Grant, L. T. Ngo, K. Stegelman, C. T. Campbell, *J. Phys. Chem. B* **2003**, *107*, 1180.

[13] H. U. Blaser, H. P. Jalett, W. Lottenbach, M. Studer, *J. Am. Chem. Soc.* **2000**, *122*, 12675.

[14] M. v. Arx, T. Mallat, A. Baiker, *Top. Catal.* **2002**, *19*, 75.

[15] M. v. Arx, T. Burgi, T. Mallat, A. Baiker, *Chem. Eur. J.* **2002**, *8*, 1430.

[16] T. Pignet, L. D. Schmidt, *Chem. Eng. Sci.* **1974**, *29*, 1123.

[17] B. Y. K. Pan, *J. Catal.* **1971**, *21*, 27.

[18] G. J. Hutchings, *Appl. Catal.* **1986**, *28*, 7.

[19] D. Hasenberg, L. D. Schmidt, *J. Catal.* **1986**, *97*, 156.

[20] D. Hasenberg, L. D. Schmidt, *J. Catal.* **1987**, *104*, 441.

[21] N. Waletzko, L. D. Schmidt, *AIChE J.* **1988**, *34*, 1146.

[22] A. Bockholt, I. S. Harding, R. M. Nix, J. *Chem. Soc., Faraday Trans.* **1997**, *93*, 3869.

[23] A. Bourane, D. Bianchi, *J. Catal.* **2001**, *202*, 34.

[24] A. Bourane, D. Bianchi, *J. Catal.* **2002**, *209*, 114.

[25] J. R. Gonzalez-Velasco, M. A. Gutierrez-Ortiz, J. L. Marc, J. A. Botas, M. P. Gonzalez-Marcos, G. Blanchard, *Ind. Eng. Chem. Res.* **2003**, *42*, 311.

[26] M. Hunger, J. Weitkamp, *Angew. Chem.* **2001**, *113*, 3040; *Angew. Chem. Int. Ed.* **2001**, *40*, 2954.

[27] S. Chinta, T. V. Choudhary, L. L. Daemen, J. Eckert, D. W. Goodman, *Angew. Chem.* **2002**, *114*, 152; *Angew. Chem. Int. Ed.* **2002**, *41*, 144.

[28] K. Eller, H. Schwarz, *Chem. Rev.* **1991**, *91*, 1121.

[29] T. F. Magnera, D. E. David, J. Michl, *J. Am. Chem. Soc.* **1987**, *109*, 936.

[30] G. S. Jackson, F. M. White, C. L. Hammill, R. J. Clark, A. G. Marshall, *J. Am. Chem. Soc.* **1997**, 119, 7567.

[31] T. Hanmura, M. Ichihashi, T. Kondow, *J. Phys. Chem. A* **2002**, *106*, 11465.

[32] K. K. Irikura, J. L. Beauchamp, *J. Am. Chem. Soc.* **1991**, *113*, 2769.

[33] K. K. Irikura, J. L. Beauchamp, *J. Phys. Chem.*. **1991**, *95*, 8344.

[34] R. Wesendrup, D. Schröder, H. Schwarz, *Angew. Chem.* **1994**, *106*, 1232; *Angew. Chem. Int. Ed. Engl.* **1994**, *33*, 1174.

[35] C. Heinemann, R. Wesendrup, H. Schwarz, *Chem. Phys. Lett.* **1995**, *239*, 75.

[36] M. Pavlov, M. R. A. Blomberg, P. E. M. Siegbahn, R. Wesendrup, C. Heinemann, H. Schwarz, *J. Phys. Chem. A* **1997**, *101*, 1567.

[37] X.-G. Zhang, R. Liyanage, P. B. Armentrout, *J. Am. Chem. Soc.* **2001**, *123*, 5563.

[38] H. Schwarz, D. Schröder, *Pure Appl. Chem.* **2000**, *72*, 2319.

[39] M. Aschi, M. Brönstrup, M. Diefenbach, J. N. Harvey, D. Schröder, H. Schwarz, *Angew. Chem.* **1998**, *110*, 858; *Angew. Chem. Int. Ed.* **1998**, *37*, 829.

[40] M. Diefenbach, M. Brönstrup, M. Aschi, D. Schröder, H. Schwarz, *J. Am. Chem. Soc.* **1999**, *121*, 10614.

[41] M. Brönstrup, D. Schröder, I. Kretzschmar, H. Schwarz, J. N. Harvey, *J. Am. Chem. Soc.* **2001**, *123*, 142.

[42] W. Ekardt, Ed., *Metal Clusters*, Wiley, Chichester, **1999**.

[43] A. Kaldor, D. M. Cox, *Pure Appl. Chem.* **1990**, *62*, 79.

[44] G. A. Somorjai, *Introduction to Surface Chemistry and Catalysis*, Wiley, New York, **1994**.

[45] M. Haruta, S. Tsubota, T. Kobayashi, H. Kageyama, M. J. Genet, B. Delmon, *J. Catal.* **1993**, *144*, 175.

[46] M. Valden, X. Lai, D. W. Goodman, *Science* **1998**, *281*, 1647.

[47] G. C. Bond, D. T. Thompson, *Catal. Rev.* **1999**, *41*, 319.

[48] J. Hagen, L. D. Socaciu, M. Elijazyfer, U. Heiz, T. M. Bernhardt, L. Wöste, *Phys. Chem. Chem. Phys.* **2002**, *4*, 1707.

[49] D. Stolcic, M. Fischer, G. Ganteför, Y. D. Kim, Q. Sun, P. Jena, *J. Am. Chem. Soc.* **2003**, *125*, 2848.

[50] H. Häkkinen, S. Abbet, A. Sanchez, U. Heiz, U. Landman, *Angew. Chem.* **2003**, *115*, 1335; *Angew. Chem. Int. Ed.* **2003**, *42*, 1297.

[51] L. D. Socaciu, J. Hagen, T. M. Bernhardt, L. Wöste, U. Heiz, H. Häkkinen, U. Landman, *J. Am. Chem. Soc.* **2003**, *125*, 10437.

[52] K. Hayek, R. Kramer, Z. Paal, *Appl. Catal. A - Gen.* **1997**, *162*, 1.

[53] A. Y. Stakheev, L. M. Kustov, *Appl. Catal. A - Gen.* **1999**, *188*, 3.

[54] E. Dias, A. T. Davies, M. D. Mantle, D. Roy, L. F. Gladden, *Chem. Eng. Sci.* **2003**, *58*, 621.

[55] J. H. Sinfelt, *Bimetallic Catalysts*, Wiley, New York, **1985**.

[56] A. Kaldor, D. M. Cox, *J. Chem. Soc., Faraday Trans.* **1990**, *86*, 2459.

[57] U. Achatz, C. Berg, S. Joos, B. S. Fox, M. K. Beyer, G. Niedner-Schatteburg, V. E. Bondybey, *Chem. Phys. Lett.* **2000**, *320*, 53.

[58] D. J. Trevor, D. M. Cox, A. Kaldor, *J. Am. Chem. Soc.* **1990**, *112*, 3742.

[59] D. J. Trevor, R. L. Whetten, D. M. Cox, A. Kaldor, *J. Am. Chem. Soc.* **1985**, *107*, 518.

[60] P. A. Hintz, K. M. Ervin, *J. Chem. Phys.* **1994**, *100*, 5715.

[61] P. A. Hintz, K. M. Ervin, *J. Chem. Phys.* **1995**, *103*, 7897.

[62] A. Grushow, K. M. Ervin, *J. Chem. Phys.* **1997**, *106*, 9580.

[63] Y. Shi, K. M. Ervin, *J. Chem. Phys.* **1998**, *108*, 1757.

[64] M. B. Comisarow, A. G. Marshall, *Chem. Phys. Lett.* **1974**, *25*, 282.

[65] M. B. Comisarow, A. G. Marshall, *Chem. Phys. Lett.* **1974**, *26*, 489.

[66] M. B. Comisarow, *Adv. Mass Spectrom.* **1978**, *7*, 1042.

[67] A. G. Marshall, C. L. Hendrickson, G. S. Jackson, *Mass Spectrom. Rev.* **1998**, *17*, 1.

[68] K. Eller, H. Schwarz, *Int. J. Mass Spectrom. Ion Processes* **1989**, *93*, 243.

[69] K. Eller, W. Zummack, H. Schwarz, *J. Am. Chem. Soc.* **1990**, *112*, 621.

[70] M. Engeser, T. Weiske, D. Schröder, H. Schwarz, *J. Phys. Chem. A* **2003**, *107*, 2855.

[71] S. Maruyama, L. R. Anderson, R. E. Smalley, *Rev. Sci. Instrum.* **1990**, *61*, 3686.

[72] C. Berg, T. Schindler, M. Kantlehner, G. Niedner-Schatteburg, V. E. Bondybey, *Chem. Phys.* **2000**, *262*, 143.

[73] B. L. Tjelta, P. B. Armentrout, *J. Phys. Chem. A* **1997**, *101*, 2064.

[74] Parameter "Source" of the instrument.

[75] The time delay between laser shot and opening of the ion gate at the inlet into the analyzer cell can also be varied.

[76] Parameter FL2 of the instrument.

[77] The given magnetic flux density $B_z = 7.05$ T only is the nominal value. The real flux density is somewhat lower because the super-conducting magnet slowly but permanently decreases in current. To account for this effect, the FT-ICR mass spectrometer has regularly to be re-calibrated.

[78] Note that only cations whose motion in z direction corresponds to translational energies $E_{trans} < V_{trap}$ are confined. More energetic ions entering the cell therefore have to be retarded by application of a repulsive potential at the inlet, see above.

[79] In the present experiments, the data size of the raw spectra is increased by zero-filling prior to Fourier transformation in order to achieve better signal-to-noise ratios. Numerical processing is performed by means of an ASPECT 3000 minicomputer.

[80] ^{192}Pt: 0.8, ^{194}Pt: 33.0, ^{195}Pt: 33.8, ^{196}Pt: 25.2, ^{198}Pt: 7.2 %.

[81] Note that the mass-selected ^{195}Pt$_m^+$ clusters are overlapped by contributions of the other isotopes according to their natural abundances. Mass-selected Pt$_2^+$ with $m/z =$ 390, for example, corresponds to a mixture mostly containing ^{195}Pt$_2^+$ and ^{194}Pt^{196}Pt$^+$ along with a trace of ^{192}Pt^{198}Pt$^+$.

[82] Besides broad-band and single-ion ejection, the instrument's software also offers an intermediate ejection mode, the so-called frequency cover. However, the generation of both well-defined and intensive covers proves to be so time-consuming that this option usually is not used.

[83] R. A. Forbes, H. F. Laukien, J. Wronka, *Int. J. Mass Spectrom. Ion Processes* **1998**, *83*, 23.

[84] More explicitly, the instrument's ejection-attenuator parameter JA is increased to 30 dB compared to 22 dB commonly used in experiments focusing on ions with $m/z <$ 200. The irradiation time of the broadband-ejection pulses is set to 100, that of the single-ejection pulses to 30 000 or 60 000 μs (note that the values entered into the instrument's interface correspond to half of the real pulse durations). With these settings, the length of the ejection procedure is limited to 5 s at maximum.

[85] The ion with m/z = 599 consists of $^{195}Pt_3CH_2^+$, $^{194}Pt^{195}Pt^{196}PtCH_2^+$, and traces of $^{192}Pt^{195}Pt^{198}PtCH_2^+$.

[86] This equivalence only holds in the absence of energetic barriers in addition to the bond-dissociation energy. In the case of ion-molecule complexes, such barriers are usually considered negligible.

[87] Added by pulsing-in or leaking-in permanently. If the preparation of the ion requires a high permanent pressure of a neutral reactant, this presumably functions as collision partner as well.

[88] P. B. Grosshans, A. G. Marshall, *Int. J. Mass Spectrom. Ion Processes* **1990**, *100*, 347.

[89] H. Sievers, H.-F. Grützmacher, P. Caravatti, *Int. J. Mass Spectrom. Ion Processes* **1996,** *157/158*, 233.

[90] H. Sievers, *PhD Thesis*, Universität Bielefeld, D 361, **1996**.

[91] J. Schwarz, *Untersuchungen zur Struktur und Reaktivität von Organometall-verbindungen in der Gasphase (PhD Thesis, Technische Universität Berlin, D 83)*, Shaker, Aachen, **1996**.

[92] Derived from JA = 30 according to ref. [91].

[93] A further decrease of the effective pressure would significantly reduce the probability of single collisions as well and thus increase the signal-to-noise ratio for the fragment ions, thus obscuring the determination of their onset.

[94] With the exception of C_3O_2, see ref. [99], all neutral reactants applied are commercial. Apart from degassing of liquids by repeated freeze-pump circles, the substrates are used without further purification.

[95] The pressure calibration relies on the measurement of the rate constants for the reactions of V^+ with O_2, Fe^+ with N_2O, and FeO^+ with H_2 which are known to correspond to $k = 2.8 \times 10^{-10}$ (G. K. Koyanagi, D. Caraiman V. Blagojevic, D. K. Bohme, *J. Phys. Chem. A* **2002**, *106*, 4590), 3.6×10^{-11} (V. Baranov, G. Javahery, A. C. Hopkinson, D. K. Bohme, *J. Am. Chem. Soc.* **1995**, *117*, 12801; J. M. C. Plane, R. J. Rollason, *J. Chem. Soc., Faraday Trans.* **1996**, *92*, 4371), and 8.8×10^{-12} cm^3 s^{-1} (ref. [98]), respectively.

[96] J. E. Bartmess, R. M. Georgiadis, *Vacuum* **1983**, *33*, 149.

[97] F. Nakao, *Vacuum* **1975**, *25*, 431.

[98] D. Schröder, H. Schwarz, D. E. Clemmer, Y.-M. Chen, P. B. Armentrout, V. I. Baranov, D. K. Böhme, *Int. J. Mass Spectrom. Ion Processes* **1997**, *161*, 175.

[99] In the case of PtC^+, its generation by reaction of kinetically excited Pt^+ with CH_4 proves inefficient. Instead, reaction of Pt^+ with carbon suboxide C_3O_2 prepared by dehydration of malonic acid (see ref. [100]), *inter alia*, gives PtC_2O^+ which after mass-selection yields PtC^+ upon CID.

[100] H. Beyer, W. Walter, *Lehrbuch der Organischen Chemie,* 22th ed., Hirzel, Stuttgart, **1991**.

[101] If, *e.g.*, only association products AB_x^+ are observed, stepwise addition can be safely inferred because the probability of termolecular processes is negligible for the low pressures applied.

[102] U. Mazurek, H. Schwarz, *ICR Kinetics, v. 3.0.1*, Technische Universität Berlin, **1998**.

[103] U. Mazurek, *PhD Thesis*, Technische Universität Berlin, D 83, **2002**.

[104] M. P. Langevin, *Ann. Chim. Phys.* **1905**, *5*, 245.

[105] G. Gioumousis, D. P. Stevenson, *J. Chem. Phys.* **1958**, *29*, 294.

[106] T. Su, W. J. Chesnavich, *J. Chem. Phys.* **1982**, *76*, 5183.

[107] T. Su, *J. Chem. Phys.* **1988**, *88*, 4102; *ibid.* **1988**, *89*, 5355.

[108] W. E. Farneth, J. I. Brauman, *J. Am. Chem. Soc.* **1976**, *98*, 7891.

[109] W. N. Olmstead, J. I. Brauman, *J. Am. Chem. Soc.* **1977**, *99*, 4219.

[110] R. Zahradník, *Acc. Chem. Res.* **1995**, *28*, 306.

[111] R. C. Dunbar in *Current Topics in Ion Chemistry and Physics*, edited by C. Y. Ng, I. Powis, Wiley, New York, **1994**, Vol. II.

[112] A system consisting of N atoms has $3N - 6$ ($3N - 5$ for linear geometries) internal degrees of freedom.

[113] R. Liyanage, M. L. Styles, R. A. J. O'Hair, P. B. Armentrout, *Int. J. Mass Spectrom.* **2003**, *227*, 47.

[114] The major dehydrogenation products ($Pt_mN_yH_{3y-2x}^+$, $m = 4, 5$ and $x = 1, 2$) observed are: $Pt_4N_5H_{11}^+$, $Pt_4N_5H_{13}^+$, $Pt_5N_3H_5^+$, $Pt_5N_3H_7^+$, $Pt_5N_4H_8^+$, $Pt_5N_4H_{10}^+$, and $Pt_5N_5H_{11}^+$.

[115] The kinetic model only accounts for consecutive reactions according to

$$Pt_mN_{y-1}H_{3(y-1)}^+ + NH_3 \rightarrow Pt_mN_yH_{3y-2x}^+ + xH_2.$$

Particularly, it does not distinguish between reactions 3.2 ($x = 0$) and 3.3 ($x = 1$) for a given number of nitrogen atoms in the product, *i.e.*, y, but considers the overall effective rate constants.

[116] U. Achatz, M. Beyer, S. Joos, B. S. Fox, G. Niedner-Schatteburg, V. E. Bondybey, *J. Phys. Chem. A* **1999**, *103*, 8200.

[117] G. Albert, C. Berg, M. Beyer, U. Achatz, S. Joos, G. Niedner-Schatteburg, V. E. Bondybey, *Chem. Phys. Lett.* **1997**, *268*, 235.

[118] See: http://webbook.nist.gov/chemistry/.

[119] S. Taylor, G. W. Lemire, Y. M. Hamrick, Z. Fu, M. D. Morse, *J. Chem. Phys.* **1988**, *89*, 5517.

[120] M. Brönstrup, *Organometallic Ion Chemistry of Iron and Platinum in the Gas Phase (PhD Thesis, Technische Universität Berlin, D 83)*, Shaker, Aachen, **2000**.

[121] Generated upon treatment of $CD_3NH_3^+Cl^-$ with wet KOH.

[122] The unfavorable pumping characteristics of CH_3NH_2 render maintenance of a constant pressure of this substrate particularly difficult.

[123] E. P. L. Hunter, S. G. Lias, *J. Phys. Chem. Ref. Data* **1998**, *27*, 413.

[124] Small fractions of cluster fragmentation observed are attributed to contamination by O_2 (see Section 3.3) although their partial origin from reaction with N_2O cannot be excluded.

[125] P. B. Armentrout, L. F. Halle, J. L. Beauchamp, *J. Chem. Phys.* **1982**, *76*, 2449.

[126] I. Kretzschmar, A. Fiedler, J. N. Harvey, D. Schröder, H. Schwarz, *J. Phys. Chem. A* **1997**, *101*, 6252.

[127] A. Stirling, *J. Am. Chem. Soc.* **2002**, *124*, 4058.

[128] J. B. Griffin, P. B. Armentrout, *J. Chem. Phys.* **1997**, *107*, 5345.

[129] K. Koszinowski, D. Schröder, H. Schwarz, R. Liyanage, P. B. Armentrout, *J. Chem. Phys.* **2002**, *117*, 10039.

[130] A. F. Holleman, E. Wiberg, *Lehrbuch der Anorganischen Chemie*, 101th ed., de Gruyter, Berlin, **1995**.

[131] G. K. Koyanagi, D. Caraiman, V. Blagojevic, D. K. Bohme, *J. Phys. Chem. A* **2002**, *106*, 4581.

[132] L. K. Verheij, M. B. Hugenschmidt, B. Poelsema, G. Comsa, *Catal. Lett.* **1991**, *9*, 195.

[133] M. Belgued, P. Pareja, A. Amariglio, H. Amariglio, *Nature* **1991**, *352*, 789.

[134] J. M. Bradley, A. Hopkinson, D. A. King, *J. Phys. Chem.* **1995**, *99*, 17032.

[135] J. M. Bradley, A. Hopkinson, D. A. King, *Surf. Sci.* **1997**, *371*, 255.

[136] G. S. Selwyn, G. T. Fujimoto, M. C. Lin, *J. Phys. Chem.* **1982**, *86*, 760.

[137] S. Y. Hwang, L. D. Schmidt, *J. Catal.* **1988**, *114*, 230.

[138] S. Y. Hwang, A. C. F. Kong, L. D. Schmidt, *J. Phys. Chem.* **1989**, *93*, 8327.

[139] W. H. Weinberg, *J. Catal.* **1973**, *28*, 459.

[140] N. R. Avery, *Surf. Sci.* **1983**, *131*, 501.

[141] M. H. Kim, J. R. Ebner, R. M. Friedman, M. A. Vannice, *J. Catal.* **2001**, *204*, 348.

[142] H. Basch, D. G. Musaev, K. Morokuma, *J. Mol. Struct. (THEOCHEM)* **2002**, *586*, 35.

[143] J. L. Gland, B. A. Sexton, G. B. Fisher, *Surf. Sci.* **1980**, *95*, 587.

[144] H. Steininger, S. Lehwald, H. Ibach, *Surf. Sci.* **1982**, *123*, 1.

[145] A. Winkler, X. Guo, H. R. Siddiqui, P. L. Hagans, J. T. Yates, Jr., *Surf. Sci.* **1988**, *201*, 419.

[146] G. Chinchen, P. Davies, R. J. Sampson in *Catalysis: Science and Technology*, edited by J. R. Anderson, M. Boudart, Springer, New York, **1987**, Vol. 8, p. 1.

[147] D. Hughes, *Chemsa* **1975**, 49.

[148] M. Brönstrup, D. Schröder, H. Schwarz, *Organometallics* **1999**, *18*, 1939.

[149] Y. A. Ranasinghe, T. J. MacMahon, B. S. Freiser, *J. Phys. Chem.* **1991**, *95*, 7721.

[150] P. B. Armentrout, M. R. Sievers, *J. Phys. Chem. A* **2003**, *107*, 4396.

[151] H. Schwarz, *Angew. Chem.* **2003**, *115*, September Issue; *Angew. Chem. Int. Ed.* **2003**, *43*, September Issue.

[152] P. B. Armentrout, B. L. Kickel in *Organometallic Ion Chemistry*, edited by B. S. Freiser, Kluwer, Dordrecht, **1996**, p. 1.

[153] J. Bigeleisen, *J. Chem. Phys.* **1949**, *17*, 675.

[154] The product ratios observed are also corrected for the incomplete deuterium content ($\geq$ 98 %) and the natural ^{13}C abundance.

[155] According to $\Delta G = - RT \ln K$, where T is the (absolute) temperature and R the gas constant.

[156] Different absolute time scales mainly result from different pressures $p(D_2)$.

[157] R. Liyanage, X.-G. Zhang, P. B. Armentrout, *J. Chem. Phys.* **2001**, *115*, 9747.

[158] E. J. Bieske, O. Dopfer, *Chem. Rev.* **2000**, *100*, 3963.

[159] N. L. Pivonka, C. Kaposta, G. v. Helden, G. Meijer, L. Wöste, D. M. Neumark, K. R. Asmis, *J. Chem. Phys.* **2002**, *117*, 6493.

[160] J. Lemaire, P. Boissel, M. Heninger, G. Mauclaire, G. Bellec, H. Mestdagh, A. Simon, S. Le Caer, J. M. Ortega, F. Glotin, P. Maitre, *Phys. Rev. Lett.* **2002**, *89*, 273002.

[161] Simple adducts $[Pt_m,C,H_2,O](H_2O)^+$ may be formed by termolecular stabilization at higher pressures. However, because of its very low operating pressures, FT-ICR mass spectrometry is not the method of choice to study such processes, which do not have direct relevance in the context of methane functionalization.

[162] With respect to the analogous mononuclear species $OPt(CO)^+$, compare: X.-G. Zhang, P. B. Armentrout, *J. Phys. Chem. A*, submitted.

[163] X.-G. Zhang, P. B. Armentrout, *Organometallics* **2001**, *20*, 4266.

[164] Calculated for $T = 0$ on the basis of $D_0(Pt^+–C) = 524 \pm 5,^{37} D_0(O–O) = 490,^{165}$ and $D_0(C–O) = 1072$ kJ mol^{-1}; see ref. [165] for further data.

[165] H. P. Huber, G. Herzberg, *Constants of Diatomic Molecules*, van Nostrand-Reinhold, New York, **1979**.

[166] P. Forzatti, L. Lietti, *Catal. Today* **1999**, *52*, 165.

[167] N. M. Ostrovskii, *Kinet. Catal.* **2001**, *42*, 326.

[168] W. Bronger, P. Müller, K. Wrzesien, *Z. Anorg. Allg. Chem.* **1997**, *623*, 362.

[169] D. Schröder, H. Schwarz, *Angew. Chem.* **1990**, *102*, 1468; *Angew. Chem. Int. Ed. Engl.* **1990**, *29*, 1433.

[170] O. Gehret, M. P. Irion, *Chem. Eur. J.* **1996**, *2*, 598.

[171] P. Jackson, J. N. Harvey, D. Schröder, H. Schwarz, *Int. J. Mass Spectrom.* **2001**, *204*, 233.

[172] M. Brönstrup, I. Kretzschmar, D. Schröder, H. Schwarz, *Helv. Chim. Acta* **1998**, *81*, 2348.

[173] K. Koszinowski, D. Schröder, H. Schwarz, unpublished results.

[174] Y. Yazawa, N. Takagi, H. Yoshida, S. Komai, A. Satsuma, T. Tanaka, S. Yoshida, T. Hattori, *Appl. Catal. A - Gen.* **2002**, *233*, 103.

[175] Y. Yazawa, H. Yoshida, T. Hattori, *Appl. Catal. A - Gen.* **2002**, *237*, 139.

[176] D. C. Koningsberger, J. de Graaf, B. L. Mojet, D. E. Ramaker, J. T. Miller, *Appl. Catal. A - Gen.* **2000**, *191*, 205.

[177] P. J. Hill, N. Adams, T. Biggs, P. Ellis, J. Hohls, S. S. Taylor, I. M. Wolff, *Mat. Sci. Eng. A* **2002**, *329*, 295.

[178] Generous gift of DEGUSSA AG.

[179] D. L. Hildenbrand, K. H. Lau, *J. Chem. Phys.* **1994**, *101*, 6076.

[180] M. W. Chase, Jr., C. A. Davies, J. R. Downey, Jr., D. J. Frurip, R. A. McDonald, A. N. Syverud, *J. Phys. Chem. Ref. Data* **1985**, *14*, Suppl. 1.

[181] By manual operation of a leak valve.

[182] C. Copéret, M. Chabanas, R. P. Saint-Arroman, J.-M. Basset, *Angew. Chem.* **2003**, *115*, 164; *Angew. Chem. Int. Ed.* **2003**, *42*, 156.

[183] R. Kojima, K. Aika, *Chem. Lett.* **2000**, 514.

[184] R. Kojima, K. Aika, *Appl. Catal.* **2001**, *215*, 149.

[185] R. Kojima, K. Aika, *Appl. Catal.* **2001**, *218*, 121.

[186] R. Kojima, K. Aika, *Appl. Catal.* **2001**, *219*, 157.

[187] C. J. H. Jacobsen, S. Dahl, B. S. Clausen, S. Bahn, A. Logadottir, J. K. Nørskov, *J. Am. Chem. Soc.* **2001**, *123*, 8404.

[188] R. Schlögl, *Angew. Chem.* **2003**, *115*, 2050; *Angew. Chem. Int. Ed.* **2003**, *42*, 2004.

[189] D. B. Jacobson, B. S. Freiser, *J. Am. Chem. Soc.* **1986**, *108*, 27.

[190] J. A. Barrow, D. J. Sordelet, M. F. Besser, C. J. Jenks, P. A. Thiel, E. F. Rexer, S. J. Riley, *J. Phys. Chem. A* **2002**, *106*, 9204.

[191] Similar methods for the removal of isobaric interferences have also been used in some other cases, see: (a) K. K. Irikura, E. H. Fowles, J. L. Beauchamp, *Anal. Chem.* **1994**, *66*, 3447; (b) G. K. Koyanagi, V. V. Lavrov, V. Baranov, D. Bandura, S. Tanner, J. W. McLaren, D. K. Bohme, *Int. J. Mass Spectrom.* **2000**, *194*, L1.

[192] See: http://www.liv.ac.uk/Chemistry/Links/links.html.

[193] D. M. Cox, R. Brickman, K. Creegan, A. Kaldor, *Z. Phys. D* **1991**, *19*, 353.

[194] F. Aguirre, J. Husband, C. J. Thompson, R. B. Metz, *Chem. Phys. Lett.* **2000**, *318*, 466.

[195] M. P. Irion, P. Schnabel, A. Selinger, *Ber. Bunsenges. Phys. Chem.* **1990**, *94*, 1291.

[196] C. Jackschath, I. Rabin, W. Schulze, *Ber. Bunsenges. Phys. Chem.* **1992**, *96*, 1200.

[197] D. Schröder, H. Schwarz, J. Hrušák, P. Pyykkö, *Inorg. Chem.* **1998**, *37*, 624.

[198] N. A. Lambropoulos, J. R. Reimers, N. S. Hush, *J. Chem. Phys.* **2002**, *116*, 10277.

[199] D. Schröder, M. Diefenbach, H. Schwarz, A. Schier, H. Schmidbaur in *Relativistic Effects in Heavy-Element Chemistry and Physics*, edited by B. A. Hess, Wiley, Chichester, **2002**, p. 245.

[200] J. A. Alonso, *Chem. Rev.* **2000**, *100*, 637.

[201] H. Häkkinen, U. Landman, *Phys. Rev. B* **2000**, *62*, R2287.

[202] B. E. Salisbury, W. T. Wallace, R. L. Whetten, *Chem. Phys.* **2000**, *262*, 131.

[203] G. Mills, M. S. Gordon, H. Metiu, *Chem. Phys. Lett.* **2002**, *359*, 493.

[204] S. Becker, G. Dietrich, H.-U. Hasse, N. Klisch, H.-J. Kluge, D. Kreisle, S. Krückeberg, M. Lindinger, K. Lützenkirchen, L. Schweikhard, H. Weidele, J. Ziegler, *Z. Phys. D* **1994**, *30*, 341.

[205] S. Gilb, P. Weis, F. Furche, R. Ahlrichs, M. M. Kappes, *J. Chem. Phys.* **2002**, *116*, 4094.

[206] D. Schröder, J. Hrušák, R. H. Hertwig, W. Koch, P. Schwerdtfeger, H. Schwarz, *Organometallics* **1995**, *14*, 312.

[207] D. M. P. Mingos, J. Yau, S. Menzer, D. J. Williams, *J. Chem. Soc., Dalton Trans.* **1995**, 319.

[208] C. Wittborn, U. Wahlgren, *Chem. Phys.* **1995**, *201*, 357.

[209] M. A. Aubart, L. H. Pignolet, *J. Am. Chem. Soc.* **1992**, *114*, 7901.

[210] M. A. Aubart, B. D. Chandler, R. A. T. Gould, D. A. Krogstad, M. F. J. Schoondergang, L. H. Pignolet, *Inorg. Chem.* **1994**, *33*, 3724.

[211] A. Cruz, E. Poulain, G. del Angel, S. Castillo, V. Berin, *Int. J. Quant. Chem.* **1998**, *67*, 399.

[212] A. Cruz, G. del Angel, E. Poulain, J. M. Martinez-Magadan, M. Castro, *Int. J. Quant. Chem.* **1999**, *75*, 699.

[213] D. Dai, K. Balasubramanian, *J. Chem. Phys.* **1994**, *100*, 4401.

[214] A. K. Chowdhury, C. L. Wilkins, *J. Am. Chem. Soc.* **1987**, *109*, 5336.

[215] Similar conclusions have been drawn from the coupling reactions of different mononuclear carbenes MCH_2^+ with NH_3; see ref. [40].

[216] With regard to the reactivity of Ag_2^+, see also: P. Sharpe, C. J. Cassady, *Chem. Phys. Lett.* **1992**, *191*, 111.

[217] M. P. Irion, A. Selinger, *Chem. Phys. Lett.* **1989**, *158*, 145.

[218] D. Walter, P. B. Armentrout, *J. Am. Chem. Soc.* **1998**, *120*, 3176.

[219] H. El Aribi, C. F. Rodriquez, T. Shoeib, Y. Ling, A. C. Hopkinson, K. W. M. Siu, *J. Phys. Chem. A* **2002**, *106*, 8798.

[220] The coinage-metal dimers already fail to activate CH_4. Alternatively generated $Au_2CH_2^+$ cannot be coupled with NH_3 either, see Section 8.1. The analogous reactivities of $Cu_2CH_2^+$ and $Ag_2CH_2^+$ cannot be probed because of lacking practicable syntheses for these ions (reaction of M_2^+ with CH_3X, X = Cl, Br, and I, does not afford the desired result).

[221] V. E. Bondybey, M. K. Beyer, *J. Phys. Chem. A* **2001**, *105*, 951.

[222] B. E. Hayden in *Vibrational Spectroscopy of Molecules on Surfaces*, edited by J. T. Yates, Jr., T. E. Madey, Plenum, New York, **1987**, Chapter 7.

[223] J. B. Griffin, P. B. Armentrout, *J. Chem. Phys.* **1997**, *107*, 5345.

[224] C. Heinemann, *Gas-Phase Ion-Molecule Chemistry: Synergy Between Experiment and Theory (PhD Thesis, Technische Universität Berlin, D 83)*, Shaker, Aachen, **1996**.

[225] P. Chen, *Angew. Chem.* **2003**, *115*, 2938; *Angew. Chem. Int. Ed.* **2003**, *42*, 2832.

Publication Index

1.) *Bond-Dissociation Energies and Structures of Cu(NO)$^+$ and Cu(NO)$_2$$^+$*
Konrad Koszinowski, Detlef Schröder, Helmut Schwarz, Max C. Holthausen,
Joachim Sauer, Hideya Koizumi, Peter B. Armentrout, *Inorg. Chem.* **2002**, *41*,
5882-5890.

2.) *Thermochemistry of small cationic iron-sulfur clusters*
Konrad Koszinowski, Detlef Schröder, Helmut Schwarz, Rohana Liyanage, Peter
B. Armentrout, *J. Chem. Phys.* **2002**, *117*, 10039-10056.

3.) *C–F Bond Activation in Fluorinated Carbonyl Compounds by Chromium
Monocations in the Gas Phase*
Ulf Mazurek, Konrad Koszinowski, Helmut Schwarz, *Organometallics* **2003**, *22*,
218-225.

4.) *Probing Cooperative Effects in Bimetallic Clusters: Indications of C–N Coupling of
CH$_4$ and NH$_3$ Mediated by the Cluster Ion PtAu$^+$ in the Gas Phase*
Konrad Koszinowski, Detlef Schröder, Helmut Schwarz, *J. Am. Chem. Soc.* **2003**,
125, 3676-3677.

5.) *Reactivity of Small Cationic Platinum Clusters*
Konrad Koszinowski, Detlef Schröder, Helmut Schwarz, *J. Phys. Chem. A* **2003**,
107, 4999-5006.

6.) *Reactions of Platinum-Carbene Clusters Pt$_n$CH$_2$$^+$ (n = 1–5) with O$_2$, CH$_4$, NH$_3$,
and H$_2$O: Coupling Processes versus Carbide Formation*
Konrad Koszinowski, Detlef Schröder, Helmut Schwarz, *Organometallics*, **2003**,
22, 3809-3819.

7.) *Additivity Effects in the Reactivities of Bimetallic Cluster Ions Pt$_m$Au$_n$$^+$*
Konrad Koszinowski, Detlef Schröder, Helmut Schwarz, *ChemPhysChem*, in press.

8.) *Formation and Reactivity of Gaseous Iron-Sulfur Clusters*
Konrad Koszinowski, Detlef Schröder, Helmut Schwarz, *Eur. J. Inorg. Chem.*, in press.

9.) *Methan-Ammoniak-Kupplung durch zweikernige bimetallische Platin-Münzmetall-Kationen PtM⁺ (M = Cu, Ag und Au)*
Konrad Koszinowski, Detlef Schröder, Helmut Schwarz, *Angew. Chem.*, accepted for publication.

10.) *C-N Coupling of Methane and Ammonia by Bimetallic Platinum-Gold Clusters*
Konrad Koszinowski, Detlef Schröder, Helmut Schwarz, *J. Am. Chem. Soc.*, submitted for publication.

Curriculum Vitae

Name	Konrad Koszinowski
Geburtsdatum	5. Januar 1976 in Hamburg
Staatsangehörigkeit	deutsch
Familienstand	ledig

1986 - 1995	Besuch der Theodor-Heuss Schule (Gymnasium), Pinneberg
Juni 1995	Erwerb der allgemeinen Hochschulreife
Juli 1995 - April 1996	Grundwehrdienst beim 61. Beobachtungsartilleriebataillon, Albersdorf (Holstein)
April 1996 - März 1998	Grundstudium der Chemie an der Universität Hamburg
Nov. 1996 - Sep. 2000	Stipendium der Studienstiftung des deutschen Volkes
Mai 1998 - September 2000	Hauptstudium der Chemie an der Ludwig-Maximilians-Universität München
März 2000 - Sep. 2000	Diplomarbeit zum Thema „Untersuchung des Gegenion-Einflusses auf die Reaktivität hochstabilisierter Carbenium- und Iminiumionen" unter Anleitung durch Prof. Dr. Herbert Mayr
Sep. 2000	Diplom-Hauptprüfung
ab Nov. 2000	Anfertigung der Dissertation „Gaseous Platinum Clusters – Versatile Models for Heterogeneous Catalysts" unter Anleitung durch Prof. Dr. Dr. h.c. Dr. h.c. Helmut Schwarz, Institut für Chemie der Technischen Universität Berlin
März 2001 - Mai 2001	Forschungsaufenthalt bei Prof. Dr. Peter B. Armentrout, University of Utah, Salt Lake City
März 2001 - Feb. 2003	Kekulé-Stipendium der Stiftung Stipendium-Fonds des Verbandes der chemischen Industrie